Preparing for the
New Jersey
GEPA
Mathematics
GRADE 8

David J. Glatzer
Supervisor of Mathematics (K–12)
West Orange Public Schools
West Orange, New Jersey

Joyce Glatzer
Supervisor of Mathematics (K–6)
West New York Public Schools
West New York, New Jersey

AMSCO SCHOOL PUBLICATIONS, INC.
315 Hudson Street, New York, N. Y. 10013

David J. Glatzer is supervisor of mathematics (K–12) for the West Orange Public Schools, West Orange, New Jersey. He has served as president of the Association of Mathematics Teachers of New Jersey (AMTNJ), member of the board of directors of the National Council of Teachers of Mathematics (NCTM), and Northeast director of the National Council of Supervisors of Mathematics (NCSM). In addition, he served as co-chair of the Mathematics Panel for New Jersey Core Course Proficiencies (New Jersey State Department of Education). He is a frequent speaker at professional conferences and has written numerous articles in professional journals including the *Arithmetic Teacher* and the *New Jersey Mathematics Teacher*. He has contributed to NCTM yearbooks and to the NCTM *Algebra for Everyone* project. In 1993, he was the recipient of the Max Sobel Outstanding Mathematics Educator Award presented by the AMTNJ. He is a co-author of the Amsco publications *Mathematics for the New Jersey HSPT* and *Preparing for the New Jersey HSPA Mathematics*.

Joyce Glatzer is supervisor of mathematics (K–6) for the West New York Public Schools, West New York, New Jersey. She has been a mathematics consultant and the former coordinator of mathematics (K–9) for the Summit Public Schools, Summit, New Jersey. She has served as president of the Association of Mathematics Teachers of New Jersey (AMTNJ) and is an active member of the NCTM. She was the 1999 recipient of the Max Sobel Outstanding Mathematics Educator Award presented by AMTNJ. She is a frequent speaker and workshop leader at professional conferences and staff development programs. She has written numerous articles in professional journals including the *New Jersey Mathematics Teacher* and the *Arithmetic Teacher*. In speaking and conducting workshops, her interests include problem solving, questioning techniques, communications, active learning with manipulatives, and use of calculators. She is a co-author of the Amsco publications *Mathematics for the New Jersey HSPT* and *Preparing for the New Jersey HSPA Mathematics*.

Reviewers
Debra Dial
Mathematics Teacher
Long Branch, New Jersey
Katherine G. Ilardi
Mathematics Educator
Oak Ridge, New Jersey
Kelly D. Mark
Mathematics Teacher
Ocean Township, New Jersey
Kelly Degnan Treshock
Mathematics Teacher
Long Branch, New Jersey

Text design by One Dot Inc.
Cover design by A Good Thing, Inc.
Composition by Maryland Comp
Line art by Monotype Composition Company

The Mathematics Reference Sheet, Generic Rubric, and some questions in the Practice Tests are reprinted courtesy of the New Jersey State Department of Education.

Please visit our Web site at: *www.amscopub.com*

When ordering this book, please specify:
either **R 310 W** *or* PREPARING FOR THE NEW JERSEY GEPA MATHEMATICS.

ISBN 1-56765-565-3

CONTENTS

Getting Started   **vii**

CLUSTER 1 — Number Sense, Concepts, and Applications   1

MACRO A   **1**
1-A-1 Estimations and Approximations   1
1-A-2 Reasonableness of Results   3
1-A-3 Significant Digits   4
Assessment Macro A   *6*

MACRO B   **8**
1-B-1 Sets of Numbers   8
1-B-2 Powers, Roots, Exponents, and Scientific Notation   11
1-B-3 Properties of Arithmetic Operations   15
1-B-4 Primes, Factors, and Multiples   17
1-B-5 Order of Operations   19
Assessment Macro B   *20*

MACRO C   **23**
1-C-1 Ratio and Proportion   23
1-C-2 Percent   26
Assessment Macro C   *31*
Assessment Cluster 1   *33*
Extra Practice: Open-Ended Questions   *37*

CLUSTER 2 — Spatial Sense and Geometry   39

MACRO A   **39**
2-A-1 Geometric Terms   39
2-A-2 Geometric Relationships   44

2-A-3 Two-Dimensional Figures 47

2-A-4 Three-Dimensional Figures and Spatial Relationships 55

Assessment Macro A 59

MACRO B **64**

2-B-1 Congruence 64

2-B-2 Similarity 66

2-B-3 Transformations 69

2-B-4 Tessellation 76

Assessment Macro B 79

MACRO C **83**

2-C-1 Perimeter and Circumference 83

2-C-2 Area 86

2-C-3 Volume 90

2-C-4 Surface Area 93

2-C-5 Standard and Non-Standard Units of Measure 96

2-C-6 Pythagorean Theorem 101

Assessment Macro C 104

Assessment Cluster 2 106

Cumulative Assessment Clusters 1 and 2 112

Extra Practice: Open-Ended Questions 115

CLUSTER 3 Data Analysis, Probability, Statistics, and Discrete Math **117**

MACRO A **117**

3-A-1 Probability of Simple Events 117

3-A-2 Probability of Compound Events 121

Assessment Macro A 125

MACRO B **127**

3-B-1 Statistical Measures 127

3-B-2 Sampling 130

3-B-3 Data Displays 131

3-B-4 Relationships Involving Data 138

3-B-5 Evaluating and Interpreting Data 140

Assessment Macro B 144

MACRO C **149**

3-C-1 Methods of Counting 149

3-C-2 Networks 159

MACRO D   **162**

3-D-1 Recursion, Iteration, and Fractals   162
3-D-2 Algorithms and Flow Charts   166
Assessment Macro C and Macro D   *170*
Assessment Cluster 3   *173*
Cumulative Assessment Clusters 1, 2, and 3   *179*
Extra Practice: Open-Ended Questions   *183*

CLUSTER 4   **Patterns, Functions, and Algebra**   **185**

MACRO A   **185**

4-A-1 Patterns   185
4-A-2 Sequences   190
4-A-3 Representations of Relationships and Patterns   194
Assessment Macro A   *197*

MACRO B   **199**

4-B-1 Expressions and Open Sentences   199
4-B-2 Linear Equations and Inequalities   204
4-B-3 Functions   208
4-B-4 Cartesian Coordinate System   213
4-B-5 Rates of Change   216
Assessment Macro B   *223*
Assessment Cluster 4   *226*
Extra Practice: Open-Ended Questions   *230*

Practice Test 1   **233**
Practice Test 2   **239**
Practice Test 3   **247**
Index   **255**

Getting Started

A. ABOUT THIS TEST

1. What is the GEPA?

The Grade Eight Proficiency Assessment (GEPA) is a standardized state assessment required of all New Jersey public school students. It includes mathematics from the four content clusters of the New Jersey Core Curriculum Content Standards. The GEPA has been developed to show whether or not you have a satisfactory level of achievement in the specified areas and to provide some idea of your potential for passing the High School Proficiency Assessment given in grade eleven. The GEPA in Mathematics is not a minimum competency or basic skills test; it is a test of your ability to do higher order thinking and to integrate topics in mathematics.

2. When do you take the GEPA?

You take the GEPA in March of eighth grade.

3. What math topics are included in the GEPA?

There are four math clusters in the GEPA.

1. Number Sense, Concepts, and Applications
2. Spatial Sense and Geometry
3. Data Analysis, Probability, Statistics, and Discrete Mathematics
4. Patterns, Functions, and Algebra

Several points need to be noted.

- Problem-solving and measurement situations exist in each cluster; as a result, there is no separate problem-solving or measurement cluster.
- Isolated computation questions do not appear on the GEPA.
- Many of the mathematics questions involve a considerable amount of reading.

4. What kinds of math questions appear on the GEPA?

There are two kinds of math questions on the test:

1. multiple-choice items
2. open-ended items

Although the multiple-choice questions on the test assess high levels of mathematical thinking, some abilities are difficult to assess with this format. Also, the format does not allow for multiple responses or for partial credit. To overcome these limitations, the test includes open-ended (free-response) questions.

Open-ended questions require you to construct your own written or graphical responses and explain these responses. The responses can be scored for different levels of mathematical understanding as well as for partial credit. More about scoring and open-ended questions follows in Section C.

5. Do you need to memorize formulas?

No. *A Mathematics Reference Sheet* is distributed to each student along with the test. This reference sheet contains any formulas you may need on the mathematics questions. The reference sheet may also contain materials such as a ruler or figures to be cut out and used in the process of solving specific questions. Refer to pages xvi–xvii for a sample of a *Mathematics Reference Sheet*.

6. Are calculators allowed on the GEPA?

Yes. You are allowed to use a calculator during the mathematics section of the GEPA. A more detailed discussion of types of calculators allowed follows in Section B.

7. How is the GEPA scored?

Multiple-choice items are worth one point each and open-ended responses are worth three points each. Refer to pages xiii-xiv for more information on the scoring of open-ended questions.

8. What is the High School Proficiency Assessment?

The High School Proficiency Assessment (HSPA) is a graduation test required of all New Jersey public school students. It is given for the first time in the spring of the junior year of high school. The HSPA covers the same math clusters as GEPA. The types of questions used on GEPA (multiple-choice and open-ended) are also used on HSPA. Calculators may be used on HSPA.

9. How can you find out more about the GEPA?

For additional information about the GEPA, ask your math teacher or guidance counselor. You can also obtain information by visiting the Web site of the New Jersey Department of Education (http://www.state.nj.us/education), opening the drop-down menu titled Overview of DOE Programs, and clicking on Assessment & Evaluation.

B. ABOUT USING A CALCULATOR

You will be allowed to use a calculator on the GEPA. The following information will help you make the most effective use of the calculator on the test.

1. On which questions should you use the calculator?

You will not need the calculator for every question on the test. With respect to calculator use, questions will fall into three categories: calculator-active, calculator-neutral, or calculator-irrelevant.

Calculator-active questions contain data that can be explored and manipulated usefully using a calculator. These questions may deal with explorations of patterns, problem solving involving guess and check, problems involving calculations with real data, or problems involving messy computation.

Calculator-neutral questions could be completed using a calculator. They may be more efficiently answered, however, by using mental math skills or simple paper-and-pencil computation. For example, the average of $-6, -7, -8, 5, 6, 7, 8, 2, 3,$ and 0 can be found more quickly mentally if you recognize that the set contains three pairs of opposites that add up to zero. In the time it takes to enter all the data into the calculator, the problem could have been solved mentally.

$\cancel{-6}, \cancel{-7}, \cancel{-8}, 5, \cancel{6}, \cancel{7}, \cancel{8}, 2, 3, 0$

$5 + 2 + 3 + 0 = 10$

$10 \div 10 = 1$

With calculator-irrelevant questions, a calculator is of no help because the solution involves no computation. For example, if a problem asks for the probability of selecting a red marble from a jar containing 2 red and 3 blue marbles, the calculator will not help answer the question.

Determining which questions to answer with the calculator is an important skill for you to develop. Be sure that you do not waste time on the test trying to use the calculator when it is not appropriate.

2. What calculator can you use?

The State Department of Education has indicated that you will be allowed to use a calculator that has at least the following functions:

a. algebraic logic (follows order of operations)
b. exponent key to use powers and roots of any degree
c. at least one memory
d. a reset button, or some other simple, straightforward way to clear all of the memory and programs

For the GEPA, use of a graphing calculator is permitted. However, calculators with QWERTY (typewriter) keyboards are not acceptable under the current guidelines.

3. What features of the calculator are you likely to need for the test?

In addition to the basic operation keys and number keys, be sure to practice using these keys:

CE/C ON/AC	clear	%	percent
M+ M− MR STO	memory	√	square root
( )	parentheses	x² yˣ	powers
+/− (−)	sign change	x!	factorial

4. What else should you consider when using the calculator on the test?

The most important thing is to be comfortable with the calculator you will be using on the test. You should be familiar with the keypad and the functions available on the calculator.

If you are using a calculator on an open-ended question, remember that it is important to show the work by writing out what you put into the calculator and the answer given.

Think before pushing the buttons. If you try to use the calculator for every question, you will waste too much time.

Be sure to estimate answers and check calculator answers for reasonableness of response.

> **Remember**
>
> Questions on the test will not tell you when to use your calculator. You must make the decision.

C. ABOUT OPEN-ENDED QUESTIONS

In addition to multiple-choice questions, the GEPA contains open-ended questions that require some writing. This section presents a variety of open-ended questions and offers suggestions for writing complete solutions. To further sharpen your skill at answering open-ended questions, try the Extra Practice questions at the end of each Cluster (pages 37, 115, 183, and 230).

1. What is an open-ended question?

An open-ended question is one in which a situation is presented, and you are asked to communicate a response. In most cases, the questions have two or more parts, and require both numerical responses and explanations. To respond completely, you may need to show calculations; make a diagram, table or graph; write sentences; or do some combination of these.

2. What might be asked in open-ended questions?

The following outline covers examples of what might be asked in these questions.

1. Make a diagram to enhance an explanation.

Example Use a diagram to show that $2\frac{1}{2} \times 2\frac{1}{2} = 6\frac{1}{4}$.

2. Complete a task (or a combination of tasks) and show your procedure.

   Example Find the area of the shaded region. Explain your procedure.

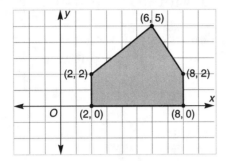

3. Give a written explanation of why a result is valid or invalid or why an approach is correct or incorrect.

   Example The average test score in a class of 20 students was 80. The average test score in a class of 30 students was 70. Mitch concluded that the average score for all 50 students was 75. He obtained the 75 by adding 80 and 70, and dividing by 2. Is his approach correct? Explain.

4. Compile a list to meet certain conditions.
   You might be asked to list numbers, dimensions, expressions, equations, etc.

   Example By looking, you should be able to tell that the average (mean) for 79, 80, 81 is 80. List three other sets of three scores that would also have an average of 80.

5. Draw a diagram to fit specific conditions.

   Example On the grid, draw three figures (a triangle, a rectangle, and a parallelogram) each with an area of 12 square units.

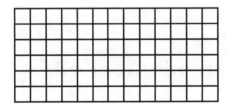

6. Describe and/or extend a pattern.

   Example Suppose the following pattern were continued.

   a. Explain how the pattern is produced.
   b. How many small squares would be in the 100th diagram?

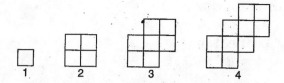

7. Indicate what will happen when a change is made in an existing situation.

   Example A rectangular prism has dimensions of 8 cm × 4 cm × 2 cm. If you double each of the three dimensions, tell what happens to the volume of the rectangular prism.

8. Complete a process involving measurement with a follow-up task.

   Example Use a ruler to determine the lengths of the sides of the accompanying figure. Label the sides with the measurements you find. Use these measurements to find the perimeter and the area of the figure. Show all work clearly.

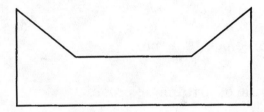

3. How are open-ended questions scored?

At the present time, the open-ended questions on the GEPA are each worth three points. (The multiple-choice items are each worth one point). Partial credit is possible on the open-ended questions. You can receive a score of 3, 2.5, 2, 1.5, 1, 0.5, or 0 for your response to an open-ended question. In questions with two or more parts, the three points a response could earn are distributed among the parts. The following examples show how the scoring could take place on different open-ended questions.

A copy of the generic rubric, which is used to develop specific rubrics for each of the open-ended items that appear on the GEPA, is presented on page xviii.

Example 1 The bottom left corner is the origin of the coordinate grid shown. The coordinates of point *A* are (3, 1), the coordinates of point *B* are (10, 7), and the coordinates of point *C* are (3, 7).

- Sketch triangle *ABC* on the coordinate grid.
- Classify the triangle according to its sides.
- Explain how you arrived at the classification.

Scoring

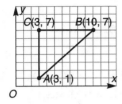

You would most likely earn one point for correctly locating the points.

You would most likely earn one point for correctly classifying the figure as a scalene triangle.

You would most likely earn one point for an acceptable explanation of the classification (that is, use of ruler, properties of right triangle).

Example 2 Karin knows that the average test score in her class of 20 students is 80. If each student in the class receives an additional 5 points in extra credit, Karin believes the class average will be 85. Is Karin correct? Explain or prove your answer to someone who disagrees with you.

Scoring You would most likely earn three points if you:

a. State that Karin is correct.

b. Support your response with an algebraic or arithmetic proof:

$$\frac{(20 \times 80) + (20 \times 5)}{20} = 85$$

c. Generalize your response in words: The sum will increase 100 points. The number of students remains constant. Hence, the change in the average is the increase divided by the number of students, or $100 \div 20 = 5$.

You would most likely receive two points if you state that Karin is correct and support your response with an arithmetic example, but do not provide a written explanation.

You would most likely receive one point if you state that Karin is correct but offer no explanation.

You would most likely receive zero points if you provide an unsatisfactory response that answers the question inappropriately.

4. What are some general guidelines for answering open-ended questions?

In answering open-ended questions, you will find the following suggestions helpful.

1. Write complete sentences.
2. Be concise, not wordy.
3. Make sure to explore different cases.
4. Answer each part of the question.
5. Make sure you answer the question that is being asked.
6. Use a diagram to enhance an explanation.
7. Label diagrams with dimensions.
8. As appropriate, give a clearly worked-out example with some explanation.
9. In problems involving estimation or approximation, make sure you round numbers *before* doing the computations.
10. Provide generalizations or written explanations as requested.
11. When using a grid, follow specific instructions for the location of the origin, axes, and so on.
12. Avoid assumptions that have no basis. For example, do not assume that a triangle is isosceles.
13. Double-check any calculations you perform within the open-ended response.
14. Be aware that a question may have more than one answer.
15. Make sure that the work you show is written down in a way that other people can understand. Label your graphs and diagrams, write neatly, and number the steps you take to solve problems.

Mathematics Reference Sheet

Use the information presented here as needed to answer questions on the mathematics portion of the Grade Eight Proficiency Assessment (GEPA).

$\pi \approx 3.14$ or $\dfrac{22}{7}$

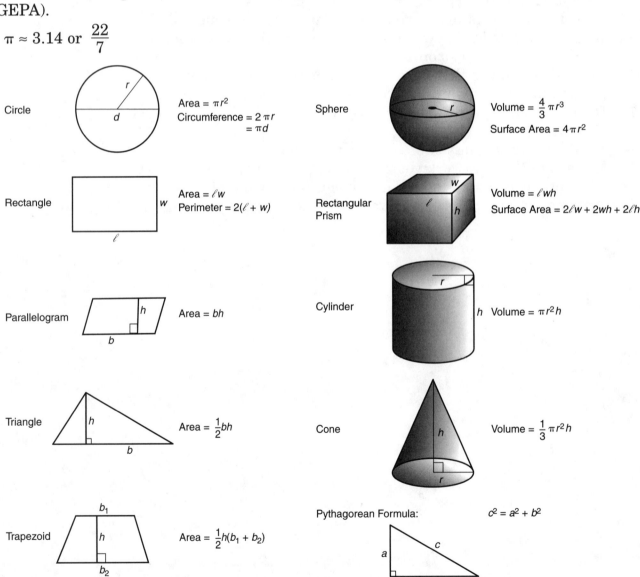

Circle

Area = πr^2
Circumference = $2\pi r$
$= \pi d$

Sphere

Volume = $\dfrac{4}{3}\pi r^3$
Surface Area = $4\pi r^2$

Rectangle

Area = ℓw
Perimeter = $2(\ell + w)$

Rectangular Prism

Volume = $\ell w h$
Surface Area = $2\ell w + 2wh + 2\ell h$

Parallelogram

Area = bh

Cylinder

Volume = $\pi r^2 h$

Triangle

Area = $\dfrac{1}{2}bh$

Cone

Volume = $\dfrac{1}{3}\pi r^2 h$

Trapezoid

Area = $\dfrac{1}{2}h(b_1 + b_2)$

Pythagorean Formula: $c^2 = a^2 + b^2$

Reprinted courtesy of the New Jersey State Department of Education.

Use the following equivalents for your calculations:

12 inches = 1 foot

10 millimeters = 1 centimeter

3 feet = 1 yard

100 centimeters = 1 meter

36 inches = 1 yard

1000 meters = 1 kilometer

5,280 feet = 1 mile

1,760 yards = 1 mile

---

16 ounces = 1 pound

1000 milligrams = 1 gram

2,000 pounds = 1 ton

100 centigrams = 1 gram

10 grams = 1 dekagram

1000 grams = 1 kilogram

---

60 seconds = 1 minute

7 days = 1 week

60 minutes = 1 hour

30 days = 1 month

24 hours = 1 day

52 weeks = 1 year

1000 watt hours = 1 kilowatt hour

---

8 fluid ounces = 1 cup

1000 milliliters (mL) = 1 liter (L)

2 cups = 1 pint

2 pints = 1 quart

4 quarts = 1 gallon

The sum of the measures of the interior angles of a triangle = 180°.

The measure of a circle is 360° or 2π radians.

Distance = rate × time   Interest = principal × rate × time

Simple Interest Formula: $A = p + prt$

Compound Interest Formula: $A = p(1 + r)^t$
  A = amount after t years r = annual interest rate
  p = principal t = number of years

The number of r-combinations of a set of size n is given by $\dfrac{n!}{(n - r)!r!}$.

The number of r-permutations of a set of n elements is given by $\dfrac{n!}{(n - r)!}$.

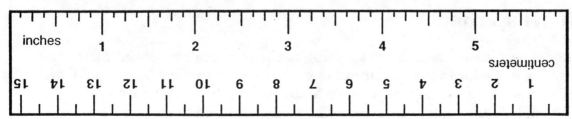

Reprinted courtesy of the New Jersey State Department of Education.

Holistic Scoring Guide for Mathematics Open-Ended Items (Generic Rubric)

3-Point Response

The response shows complete understanding of the problem's essential mathematical concepts. The student executes procedures completely and gives relevant responses to all parts of the task. The response contains few minor errors, if any. The response contains a clear, effective explanation detailing how the problem was solved so that the reader does not need to infer how and why decisions were made.

2-Point Response

The response shows nearly complete understanding of the problem's essential mathematical concepts. The student executes nearly all procedures and gives relevant responses to most parts of the task. The response may have minor errors. The explanation detailing how the problem was solved may not be clear, causing the reader to make some inferences.

1-Point Response

The response shows limited understanding of the problem's essential mathematical concepts. The response and procedures may be incomplete and/or may contain major errors. An incomplete explanation of how the problem was solved may contribute to questions as to how and why decisions were made.

0-Point Response

The response shows insufficient understanding of the problem's essential mathematical concepts. The procedures, if any, contain major errors. There may be no explanation of the solution or the reader may not be able to understand the explanation. The reader may not be able to understand how and why decisions were made.

Reprinted courtesy of the New Jersey State Department of Education.

CLUSTER 1

Number Sense, Concepts, and Applications

MACRO A

Make appropriate estimations and approximations.

1 A 1 Estimations and Approximations

An **estimate** tells you *about how much*. Estimates of number values can be used to approximate answers.

To estimate a number value:

1. Consider the use of rounded numbers.
   Use 100×24 for 97×24.
2. Consider the use of easier numbers.
   Use 20% or $\frac{1}{5}$ of 25 for 19% of 25.
3. Consider the use of benchmarks, such as with numbers
   close to $\frac{1}{2}$.
   Since $\frac{4}{10}$ is less than $\frac{1}{2}$, $\frac{4}{10}$ of 18 is less than 9.
4. Consider number relationships.
   Since $20\% = 2 \times 10\%$ and 10% of 82 is 8.2, 20% of 82 is 16.4.

Model Problem

1. Estimate 19% of $24.00.

Solution   Use easier numbers and number relationships.

Use 20% of $25.00 for 19% of $24.00.

Since 20% = 2 × 10% and 10% of 25 is 2.5, 20% of 25 is 5.

Answer   19% of $24.00 is approximately $5.00.

2. Between which two integers does $\sqrt{77}$ fall?

Solution   Recognize that $\sqrt{77}$ is not a perfect square number. Check the square roots of some perfect square numbers to see where 77 falls.

Since $8^2 = 64$, $\sqrt{64} = 8$.

Since $9^2 = 81$, $\sqrt{81} = 9$.

Answer $\sqrt{77}$ is between 8 and 9.

3. According to a county survey, $\frac{5}{8}$ of teachers read the daily newspaper. There are 243 teachers in the county. Approximately how many read the daily newspaper?

Solution   Use $\frac{1}{2}$ as a benchmark and use rounding.

Since $\frac{5}{8}$ is greater than $\frac{1}{2}$, more than $\frac{1}{2}$ of the teachers read the daily newspaper.

Round 243 to 250. Half of 250 is 125.

Answer   Approximately 125 teachers read the daily newspaper.

PRACTICE

1. Between which two integers does $\sqrt{110}$ fall?

   A. 9 and 10   B. 10 and 11
   C. 11 and 12   D. 100 and 110

2. 31π is between

   A. 80 and 90   B. 90 and 100
   C. 100 and 110   D. 110 and 120

3. Which of the following is NOT equal to the other numbers?

   A. $\frac{1.5}{4.5}$ B. $\frac{8}{24}$

   C. $\sqrt{\frac{1}{9}}$ D. $\sqrt{\frac{2}{6}}$

4. Estimate 32% of 90.

5. $\sqrt{200}$ is closest to what integer?

6. Explain why 1,000 is a good estimate for:

$$\frac{51.9 \times 102}{4.9}.$$

7. From the set of fractions $\frac{3}{4}, \frac{5}{11}, \frac{14}{27}, \frac{1}{5}$, indicate which ones have a value close to $\frac{1}{2}$. Explain the relationship that must exist between the numerator and denominator for a fraction to be close to $\frac{1}{2}$.

8. Which of the following is NOT a correct statement?

   A. 97% of 97 is less than 97.
   B. 83% of 50 is greater than 50.
   C. 105% of 80 is greater than 80.
   D. 100% of 80 is equal to 80.

---

1 A 2 Reasonableness of Results

In problem-solving situations, it is important to determine if numerical results are reasonable.

To judge the reasonableness of results, ask yourself:

1. Does the size of the answer make sense? (Should the answer be close to 100, to 1,000, or to 10,000?)
2. Does the answer fit the context or conditions of the problem?

Model Problem

The stereo that Maya wants to buy costs $420. Since the sales tax rate is 7%, Maya thinks that she will have to pay an additional $294 in tax. Is her thinking correct?

Solution $294 for sales tax would not be reasonable. $294 is greater than 50% of the selling price.

Answer Maya's thinking is incorrect. She calculated 70% instead of 7%.

1. Which of the following is NOT a correct statement?

   A. For a specific purchase, 7% sales tax could be $7,000.00
   B. The average of four even numbers must be even.
   C. The square root of a three-digit number must be less than 35.
   D. The product $0.9 \times 0.8 \times 0.7$ must be less than any one of the three given factors.

2. Which of the following would be a reasonable value for the percent of the figure that is shaded?

   A. 40%   B. 10%
   C. 25%   D. 75%

3. Dimitri used a calculator to obtain the mean for 1.25, 3.81, 0.9, 1.82, 2.01. He came up with a mean of 0.85. Is this a reasonable result? Explain.

4. In adding the fractions $\frac{1}{2} + \frac{2}{3}$, Sally came up with $\frac{3}{5}$ by adding the numerators and adding the denominators. Explain why this process results in an unreasonable answer.

5. Patricia determined the average of the following grades to be 70.

   64, 72, 74, 78, 88

   Explain why 70 would NOT be a reasonable average.

---

1 A 3 Significant Digits

Significant digits are the digits of a measurement that represent meaningful data. The number of significant digits shows how **accurate**, or close to the true value, a measurement is. The more significant digits a measurement has, the more accurate it is.

Example
A zookeeper records that a python is 16 feet long. A measurement such as 16 feet is accurate to two significant digits. It is understood that the final digit, the 6, could be wrong, so the last significant digit in any measurement is always the estimated digit. The true length of the python is within one foot of the recorded measurement, that is, in the range of 15.5 to 16.5 feet (16 ± 0.5).

If the zookeeper had recorded the python's length as 16.4 feet, the measurement would have three significant digits. The estimated digit would be the tenths digit. The true length of the python would then be within one-tenth of one foot of the recorded measurement. It would be within a smaller range of 16.35 to 16.45 feet (16.4 ± 0.05).

These rules are used to identify significant digits:

- All nonzero digits are significant.

Zero is significant in the following situations.

- When zero is used as a placeholder between two nonzero digits; for example, 105 or 1,004
- When final zeros appear after the decimal point; for example, 4.20 or 15.300
- When the final zero (or zeros) in whole numbers are marked to show they are significant; for example, 2<u>00</u> or 5<u>000</u>
- Initial zeros are never significant; for example, 0.0056 has only two significant digits, 5 and 6.

Model Problem

1. What are the significant digits in each number?

a.	584	3 significant digits 5, 8, 4
b.	46.290	5 significant digits 4, 6, 2, 9, 0
c.	0.400	3 significant digits 4, 0, 0
d.	1,500	2 significant digits 1, 5
e.	1,5<u>00</u>	4 significant digits 1, 5, 0, 0
f.	0.0871	3 significant digits 8, 7, 1

2. Give the possible range of the actual measurement for each.

   a. 140 cm   b. 14<u>0</u> cm

Solution

a. There are two significant digits 1 and 4. Since 4 is the estimated digit and indicates tens, the actual measurement is 140 ± 5 cm or the actual measurement is within the range of 135 to 145 cm.

b. There are three significant digits, 1, 4, and 0. Since 0 is the estimated digit and indicates ones, the actual measurement is 140 ± 0.5 cm or the actual measurement is within the range of 139.5 to 140.5 cm.

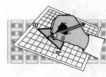

PRACTICE

1. How many significant digits are there in the measurement 1,708.30 units?

   A. 3   B. 4
   C. 5   D. 6

2. The length of a machine part is required to be 3.0 ± 0.05 cm. Which of the following lengths would NOT be acceptable?

   A. 2.98 cm   B. 3.04 cm
   C. 2.93 cm   D. 3.0 cm

3. The mass of a piece of quartz rock is reported as 39.1 g. The actual measurement is between

   A. 39.0 g and 39.2 g
   B. 39.05 g and 39.15 g
   C. 38.6 g and 39.6 g
   D. 38.10 g and 40.10 g

4. Which number has exactly four significant digits?

   A. 4,008.19   B. 27<u>00</u>
   C. 136.02   D. 0.0065

5. Indicate how many digits are significant in each number. Identify these digits.

   a. 410,600   b. 0.008
   c. 209.325   d. 15.200
   e. 5,<u>000</u>

6. a. Write a five-digit number that has only three significant digits.
   b. Show how the number you wrote can be changed to make all five digits significant.

Assessment Macro A

1. Which of the following is NOT a correct statement?

   A. $\frac{2}{5}$ of 46 is less than 23.

   B. $\frac{7}{8}$ of 87 is greater than 87.

   C. $\frac{4}{3}$ of 122 is greater than 122.

   D. $\frac{1}{9}$ of 88 is greater than 8.8.

2. Which of the following would be a reasonable value for the percent of the figure that is shaded?

   A. 20%   B. 30%
   C. 70%   D. 80%

3. Which of the following procedures would result in the best estimate for $\frac{3}{11}$ of 47?

   A. $\frac{3}{10}$ of 50   B. 30% of 50

   C. $\frac{1}{4}$ of 48   D. $\frac{1}{5}$ of 50

4. Which of the following characteristics is most reasonable for the solution to this equation?

$$2x + 7 = 4$$

   A. x is greater than 0
   B. x is a multiple of 2
   C. x is less than 0
   D. $x = 0$

5. An orangutan's weight is recorded as 113.5 lb. In what range will the true weight of the orangutan fall?

   A. 112 to 114 lb
   B. 113 to 114 lb
   C. 113.4 to 113.6 lb
   D. 113.45 to 113.55 lb

6. Show two different ways to estimate 32% of 66. Explain each process.

7. A three-digit number is multiplied by another three-digit number. Explain why the product of the two numbers may have five or six digits, but not seven digits.

8. In using the y^x key on his calculator, Emilio came up with 5.85 for a root of 200. Is the displayed value more reasonable for $\sqrt{200}$ or $\sqrt[3]{200}$? Explain your answer.

9. Estimate 19% of 500. Explain your process.

10. Estimate 198% of 4,003. Explain your process.

11. Explain why 60 is a good estimate for 19π.

12. Explain why 3 is a good estimate for $\sqrt[3]{30}$.

Open-Ended Questions

13. Ticket sales for the movie *The Matrix Reloaded* grossed $209.5 million in its first weekend released, making it the number one film for that weekend. It fell to second place during its second weekend, grossing only $45.6 million.

    a. Approximately what percent of the first weekend's gross did the film make the second weekend? Show how you arrived at your approximation.

    b. If 25% of the total ticket sales for the two weekends were child tickets, how much of the gross came from the sale of child tickets? Show your work.

14. Justin plays basketball for the Shooting Stars. He makes about 47% of his foul shots.

    a. If he shoots 97 foul shots, approximately how many foul shots will he make? Show how you got your answer. Show how you arrived at your approximation.

    b. If he takes 119 foul shots, approximately how many shots would he have to make to improve his shooting percent to 75%? Show how you arrived at your approximation.

15. A newspaper headline reported that approximately 40,000 people attended a concert. Ellen thinks that means that the maximum number of people who could have attended was 39,999.

    a. Do you agree with Ellen? Explain why or why not.

    b. The promoters claimed to have grossed approximately $800,000 on ticket sales. Assuming all tickets were the same price, about how much was the cost of a ticket? Explain.

16. a. Record the length of the key in the figure below. How many significant digits does your measurement have?

    b. Based on the measurement you recorded, in what range does the actual length of the key fall?

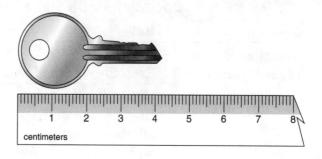

Understand numbers, our numeration system, and their applications in real-world situations.

1 B 1 Sets of Numbers

Integers are the set of whole numbers and their opposites {... ,−2, −1, 0, 1, 2, ...}. An integer has a sign (+, −), which indicates whether it is to the right or left of zero on a number line, and a **magnitude** (distance from zero on the number line). −2 is located two units to the left of zero on the number line, while +2 is located two units to the right of zero on the number line. The magnitude of a number is measured by its *absolute value*.

The **absolute value** (symbol $| \ |$) of a nonzero number is always positive. The absolute value of zero is zero: $|0| = 0$.

We describe the absolute value of a nonzero number:

1. **as a distance**. Since the point representing +3 is a distance of 3 units from zero on a number line and the point representing −3 is also a distance of 3 units from zero, the absolute value of both +3 and −3 is 3.

   In symbols:

   $$|+3| = 3 \qquad |-3| = 3 \qquad |+3| = |-3| = 3$$

2. **using opposites**. For every pair of opposite numbers, the greater number of the pair is the absolute value of both numbers. Since +3 is the greater of the pair of opposites +3 and −3, the absolute value of both +3 and −3 is 3.

   In symbols:

   $$|+3| = |-3| = +3 \text{ or } 3$$

Operations with Signed Numbers		
Operation	**Rule**	**Examples**
Addition of Signed Numbers	*Like Signs* Keep the sign, and add.	$(+3) + (+2) = +5$ $(-3) + (-2) = -5$
	Unlike Signs Use the sign of the number with the larger absolute value, and subtract.	$(+3) + (-2) = +1$ $(-3) + (+2) = -1$

Operation	Rule	Examples
Subtraction of Signed Numbers	Change the problem to addition by using the opposite of the second number.	$(+3) - (+2) = (+3) + (-2) = +1$ $(-3) - (-2) = (-3) + (+2) = -1$ $(+3) - (-2) = (+3) + (+2) = +5$ $(-3) - (+2) = (-3) + (-2) = -5$
Multiplication and Division of Signed Numbers	*Like Signs* The answer is positive.	$(+10) \times (+2) = +20 \quad (+10) \div (+2) = +5$ $(-10) \times (-2) = +20 \quad (-10) \div (-2) = +5$
	Unlike Signs The answer is negative.	$(+10) \times (-2) = -20 \quad (+10) \div (-2) = -5$ $(-10) \times (+2) = -20 \quad (-10) \div (+2) = -5$

Rational Numbers

A **rational number** is any number that can be expressed in the form $\frac{a}{b}$, with $b \neq 0$. This definition includes integers, fractions, and decimals.

Rational numbers may be compared and placed in order.

A given rational number can be expressed in different forms:

$$\frac{1}{2} = 0.5 = \frac{5}{10} = 50\% = -\left(-\frac{1}{2}\right)$$

 Model Problem

1. Which of the following is NOT equal to the other three?

   A. $\frac{2}{10}$

   B. $1.5 \div 7.5$

   C. $1 \div \frac{1}{5}$

   D. $\sqrt{\frac{1}{25}}$

Solution   A: $\frac{2}{10} = \frac{1}{5}$

B: $1.5 \div 7.5 = 0.2$ or $\frac{1}{5}$

C: $1 \div \frac{1}{5} = 1 \times 5 = 5$

D: $\sqrt{\frac{1}{25}} = \frac{1}{5}$

Answer   Choice C is not equivalent to the others.

2. Place the following rational numbers in order from LEAST to GREATEST.

$$-\frac{5}{2}, \frac{2}{5}, -3.2, 0.35$$

Solution As you consider placement of rational numbers on a number line, numbers to the left are smaller than numbers to the right.

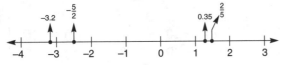

Answer $-3.2, -\frac{5}{2}, 0.35, \frac{2}{5}$

3. Which set of numbers contains all of the following?

$$0.33..., \frac{7}{11}, -6, -0.25$$

A. rational numbers
B. integers
C. irrational numbers
D. terminating decimals

Solution The rational numbers contain integers, repeating and terminating decimals, and fractions. Therefore, all four numbers listed are rational numbers.

Although -6 is an integer, the other three numbers are not integers. None of the numbers are irrational.

Although -0.25 is a terminating decimal, 0.33... is not a terminating decimal.

The fraction $\frac{7}{11}$ can be expressed as a repeating decimal, not a terminating decimal.

Answer A

PRACTICE

1. Write these fractions in order from LEAST to GREATEST.

$$\frac{1}{3}, \frac{5}{6}, \frac{2}{9}$$

A. $\frac{1}{3}, \frac{2}{9}, \frac{5}{6}$ B. $\frac{2}{9}, \frac{1}{3}, \frac{5}{6}$

C. $\frac{5}{6}, \frac{1}{3}, \frac{2}{9}$ D. $\frac{5}{6}, \frac{2}{9}, \frac{1}{3}$

2. Arrange the following numbers in order from LEAST to GREATEST.

$$\frac{1}{3}, \frac{2}{5}, 0.6, 0.125$$

A. $0.125, 0.6, \frac{1}{3}, \frac{2}{5}$

B. $0.125, \frac{1}{3}, \frac{2}{5}, 0.6$

C. $\frac{1}{3}, \frac{2}{5}, 0.125, 0.6$

D. $0.125, \frac{1}{3}, 0.6, \frac{2}{5}$

3. For a series of eight football plays, the Westwood Wolverines had the following results:

+4 yards, +3 yards, +9 yards,
−4 yards, −5 yards, +10 yards,
−5 yards, +8 yards.

What is the average yardage for this series of plays?

A. 6   B. 20
C. 2.5   D. −2.5

4. Which of the following is NOT true?

  A. The reciprocal of a negative number is negative.
  B. The sum of five positive numbers and three negative numbers must be positive.
  C. The product of an odd number of negative numbers is negative.
  D. The sum of six negative numbers must be negative.

5. Which point on the number line could represent the product of the numbers P, Q, and R?

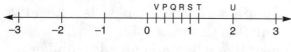

  A. S B. T
  C. U D. V

6. Two-thirds of the way into the baseball season, the Rockets have won 70 games. At this rate, how many games will they win for the entire season? Explain your procedure.

7. Which of the following is a way to find $\frac{3}{4}$ of a number?

  A. Multiply the number by $\frac{4}{3}$.
  B. Multiply the number by 3 and divide the result by 4.
  C. Divide the number by 3 and multiply the result by 4.
  D. Divide the number by 4 and divide the result by 3.

8. Mark was asked to multiply 24.3 times 0.21. He computed the product to be 51.03. Without doing the computation, explain why Mark's answer could NOT be correct.

9. Which of the two division problems, $6.25 \div 0.25$ and $6.25 \div 2.5$, would result in a greater quotient? Explain your reasoning.

10. If you know that $\frac{1}{8} = 0.125$, how can you use that fact to find the value of $\frac{7}{8}$?

11. Explain why the decimal 0.33 is not exactly equivalent to the fraction $\frac{1}{3}$.

12. Selene thinks that $4 \times 5\frac{1}{4} = 20\frac{1}{4}$. Explain why this is incorrect.

---

1 B 2 Powers, Roots, Exponents, and Scientific Notation

Exponents

An **exponent** tells how many times a base is used as a factor.

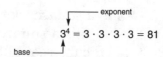

	Law	Example
Multiplication	$x^a \cdot x^b = x^{a+b}$	$2^4 \cdot 2^3 = 2^{4+3} = 2^7$
Division	$x^a \div x^b = x^{a-b}$	$3^5 \div 3^3 = 3^{5-3} = 3^2$
Powers	$(x^a)^b = x^{ab}$	$(5^2)^3 = 5^{2\times3} = 5^6$
Zero Exponent	$x^0 = 1$	$6^0 = 1$
Negative Exponent	$x^{-a} = \dfrac{1}{x^a}$	$7^{-3} = \dfrac{1}{7^3}$

Laws of Exponents ($x \neq 0$)

Model Problem

1. In simplifying $\dfrac{5^5 \times 5^3}{5^6}$, Jessica is tempted to use her calculator to find 5^5, 5^3, and 5^6 individually prior to doing the multiplication and division. Show how Jessica could use the laws of exponents to arrive at the result in a more efficient manner.

Solution $5^5 \times 5^3 = 5^{5+3} = 5^8$

$5^8 \div 5^6 = 5^{8-6} = 5^2 = 25$

2. Which of the three expressions is equivalent to the expression 16^3?

I. 2^{12} II. 4^6 III. $2^3 \cdot 8^3$

A. I, II, and III   B. only I and II
C. only II and III   D. only I

Solution   Since 16 is equivalent to 2^4, then 16^3 is equivalent to $(2^4)^3$ or 2^{12}. Thus, I is an equivalent expression.

Since 16 is equivalent to 4^2, then

16^3 is equivalent to $(4^2)^3$ or 4^6. Thus, II is an equivalent expression.

$2^3 \cdot 8^3$ is equivalent to $(2 \cdot 8)^3$ or 16^3. Thus, III is an equivalent expression.

Answer   A

3. Uri argued that the value of $(-3)^n$ will always be negative. Karen disagreed, saying that it depends on the value of n. Who is correct?

Solution   Since the base is a negative number, the value of $(-3)^n$ will be negative when n is an odd number and positive when n is an even number. Observe the pattern:

$(-3)^1 = -3$
$(-3)^2 = -3 \times -3 = 9$
$(-3)^3 = -3 \times -3 \times -3 = -27$
$(-3)^4 = -3 \times -3 \times -3 \times -3 = 81$
$(-3)^5 = -3 \times -3 \times -3 \times -3 \times -3$
$\qquad = -243$

Answer   Karen is correct.

Roots

A **root** is the inverse of a power.

If $b^2 = a$, then b is a square root of a. $b = \sqrt{a}$
If $b^3 = a$, then b is a cube root of a. $b = \sqrt[3]{a}$

Examples $5^2 = 25 \to 5$ is a square root of 25. $5 = \sqrt{25}$
$(-5)^2 = 25 \to -5$ is a square root of 25. $-5 = \sqrt{25}$
$2^3 = 8 \to 2$ is a cube root of 8. $2 = \sqrt[3]{8}$

Model Problem

If $x^2 = 4$, find two values of x that make the statement true.

Solution   Since $2^2 = 4$ and $(-2)^2 = 4$, 2 and -2 are two values that satisfy the equation. Therefore, 2 and -2 are each square roots of 4.

Scientific Notation

Scientists often work with very small and very large numbers. Scientific notation makes working with such numbers much easier. A number in **scientific notation** is expressed as a product of two factors:

(first factor is between 1 and 10) × (second factor is a power of 10).

To make the first factor, move the decimal to the right or left the necessary number of places. This movement of the decimal determines what power of ten must be used for the second factor. Use the number of moves as the exponent. If the decimal moves left, the exponent will be positive. If the decimal moves right, the exponent will be negative.

Examples

Express 4,300,000 in scientific notation.

Move the decimal 6 places to the left to make the first factor: 4.3. The second factor is then 10^6. $4,300,000 = 4.3 \times 10^6$.

Express 0.000043 in scientific notation.

Move the decimal 5 places to the right to make the first factor: 4.3. The second factor is then 10^{-5}. $0.000043 = 4.3 \times 10^{-5}$.

 Model Problem

Calculate the product of $84,500,000 \times 0.0000045$ by using scientific notation for each factor.

Solution

$84,500,000 \times 0.0000045$	
$(8.45 \times 10^7) \times (4.5 \times 10^{-6})$	Express both numbers in scientific notation.
$(8.45 \times 4.5) \times (10^7 \times 10^{-6})$	Use the commutative property to reorganize.
$(8.45 \times 4.5) \times 10^1$	Multiply the powers of 10 by adding the exponents.
38.025×10^1	Multiply the first factors.
38.025×10	Simplify.
380.25	

Answer 380.25

PRACTICE

1. Juan has \$2 invested. If his money doubles each day, which expression shows how much money he has after 5 days?

   A. 2×5
   B. $2 + 2 + 2 + 2 + 2$
   C. 2^5
   D. 5^2

2. 5,930,000 expressed in scientific notation is

   A. 5.93×10^{-6} B. 5.93×10^7
   C. 5.93×10^6 D. 59.3×10^{-5}

3. What is the perimeter of a square whose area is 49 square feet?

   A. 7 feet   B. 14 feet
   C. 21 feet   D. 28 feet

4. If the volume of a cube is 512 cubic inches, which of the following represents the length of an edge of the cube?

   A. $512 \div 3$ B. $\sqrt{512}$
   C. 512^3 D. $\sqrt[3]{512}$

5. Which of the following is NOT correct?

   A. $3^8 = (3^4)^2$ B. $3^4 \times 3^4 = 3^{16}$
   C. $3^4 \times 3^4 = 3^8$ D. $3^{10} \div 3^5 = 3^5$

6. Which of the following symbols could you write in the blank to produce a true statement?

   $$2.5 \times 10^{-4} \underline{\quad} 3.4 \times 10^{-3}$$

   A. $>$
   B. $<$
   C. $=$
   D. The comparison cannot be determined without more information.

7. Which of the following would NOT represent a large number?

   A. 2.279×10^8 B. 3.92×10^{-6}
   C. 5.9×10^{24} D. 7.37×10^{22}

8. Which of the following is NOT equivalent to 5^{-2}?

   A. -25 B. $\dfrac{1}{25}$

   C. 0.04 D. $\left(\dfrac{1}{5}\right)^2$

9. Explain the difference between $a^3 \times a^5$ and $(a^3)^5$.

10. Show how scientific notation can be used to simplify the amount of computation in the following problem:

    $$52,000 \times 1,200,000$$

1 B 3 Properties of Arithmetic Operations

Addition and multiplication have these special properties for all numbers a, b, and c.

Property	Addition	Multiplication
Commutative Property	The order in which you add two numbers does not change the sum. $6 + 5 = 5 + 6$ $a + b = b + a$	The order in which you multiply two numbers does not change the product. $8 \times 4 = 4 \times 8$ $a \times b = b \times a$
Associative Property	The way you group numbers does not change the sum. $(11 + 5) + 4 = 11 + (5 + 4)$ $(a + b) + c = a + (b + c)$	The way you group numbers does not change the product. $(3 \times 8) \times 7 = 3 \times (8 \times 7)$ $(a \times b) \times c = a \times (b \times c)$
Identity Property	If you add 0 to any number, that number remains the same. $31 + 0 = 0 + 31 = 31$ $a + 0 = 0 + a = a$	If you multiply any number by 1, that number remains the same. $97 \times 1 = 1 \times 97 = 97$ $a \times 1 = 1 \times a = a$
Zero Property		If you multiply any number by 0, the product is 0. $416 \times 0 = 0 \times 416 = 0$ $a \times 0 = 0 \times a = 0$
Inverse Property	The sum of a number and its inverse is 0. $14 + (-14) = 0$ $a + (-a) = 0$	The product of a number and its inverse is 1. $4 \times \dfrac{1}{4} = 1$ $a \times \dfrac{1}{a} = 1, a \neq 0$
Distributive Property of Multiplication over Addition or Subtraction	When you *distribute* multiplication over addition or subtraction, the result is the same whether you multiply first and then add (or subtract) or whether you add (or subtract) and then multiply. $2(3 + 5) = (2 \times 3) + (2 \times 5)$ $\qquad$ $9(6 - 4) = (9 \times 6) - (9 \times 4)$ $a(b + c) = (a \times b) + (a \times c)$ $\qquad$ $a(b - c) = (a \times b) - (a \times c)$	

Model Problem

Which property has been applied to allow the product of 15(98) to be computed mentally?

$$15(98) = 15(100-2) = 1,500 - 30 = 1,470$$

A. associative property of multiplication
B. distributive property
C. associative property of addition
D. inverse property of addition

Solution   98 is rewritten as the equivalent expression $100 - 2$ and then the 15 is distributed over the 100 and the 2. The difference can easily be found. The key is to rewrite the problem using numbers that make mental computation simple.

Answer   B

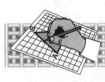

PRACTICE

1. The property that has been applied to help simplify the expression $80 + 276 + 20 = 80 + 20 + 276$ is

   A. associative property of addition
   B. commutative property of addition
   C. distributive property
   D. inverse property of addition

2. The reciprocal of $-3\frac{1}{6}$ is:

   A. $-\dfrac{19}{6}$ B. $-\dfrac{6}{19}$

   C. $\dfrac{6}{19}$ D. $\dfrac{19}{6}$

3. Which property explains why the two formulas given below for the perimeter of a rectangle are equivalent?

   $$P = 2(l + w)$$
   $$P = 2l + 2w$$

   A. associative property of multiplication
   B. distributive property
   C. commutative property of multiplication
   D. associative property of addition

4. Which of the following expressions could be rewritten using the distributive property?

   A. $a + (b - c)$ B. $a(b - c)$
   C. $a - (b \div c)$ D. $ab - c$

5. Deanne bought six T-shirts at $8.95 each. She figured out her total purchase by using this shortcut:

   $$6 \times \$9 - 6 \times \$0.05 = \$53.70.$$

   Her shortcut is an illustration of which property?

1 B 4 Primes, Factors, and Multiples

Numbers can be classified as *prime* or *composite*. A **prime number** is a whole number greater than 1 with exactly two factors, 1 and the number. The first six prime numbers are:

$$2, 3, 5, 7, 11, 13, \ldots$$

A **composite number** is a whole number greater than 1 with more than two factors. The first six composite numbers are:

$$4, 6, 8, 9, 10, 12, \ldots$$

One number is a **factor** of another number if it evenly divides that number. Thus, 6 is a factor of 18, since $18 \div 6 = 3$. But 8 is not a factor of 18, since $18 \div 8$ is not equal to a whole number.

One number is **divisible** by another if the remainder is 0 when you divide. For example, 45 is divisible by 9 because $45 \div 9 = 5$, with no remainder.

Two or more numbers may share common factors. The largest shared common factor is called the **greatest common factor (GCF)**. For example, the GCF of 18 and 48 is 6.

A **multiple** of a number is the product of that number and any other whole number. For example, 20 is a multiple of 5, since $4 \times 5 = 20$. But 18 is not a multiple of 4, since no whole number times 4 equals 18.

Two or more numbers may share a common multiple. The smallest shared common multiple is called the **least common multiple (LCM)**. The LCM of 12 and 15 is 60.

 Model Problem

1. Mr. Smart likes to describe his age in the following way: "If you divide my age by 7, the remainder is 1. If you divide my age by 2, the remainder is 1. If you divide my age by 3, the remainder is 1. My age is not divisible by 5 and is less than 100." How old is Mr. Smart?

Solution   If Mr. Smart's age is divisible by 7, 2, and 3, his age would be

$$7 \cdot 3 \cdot 2 = 42.$$

To obtain a remainder of 1 on each division, his age must be $42 + 1$ or 43.

No other possible age fits the conditions.

Answer   Mr. Smart is 43.

2. The members of the Decorating Com-mittee for a school dance have 36 red carnations, 48 white carnations, and 60 pink carnations. They want to form identical centerpieces, using all of the carnations, so that each one has the same combination of colors as the other centerpieces. What is the largest number of centerpieces they can make?

Solution   Factor each number.

red: 36: (1, 2, 3, 4, 6, 9, 12, 18, 36)
white: 48: (1, 2, 3, 4, 6, 8, 12, 16, 24, 48)
pink: 60: (1, 2, 3, 4, 5, 6, 10, 12, 15, 20, 30, 60)
common factors: (1, 2, 3, 4, 6, 12)

Answer   GCF = 12

3. What is the least number of pencils that could be packaged evenly in groups of 8 pencils OR groups of 12 pencils?

Solution

packages of 8 could hold:
(8, 16, 24, 32, 40, 48,...)
packages of 12 could hold:
(12, 24, 36, 48, 60,...)

Answer The least number possible is 24.

PRACTICE

1. Which of the numbers below has all of the following characteristics?

- It is a multiple of 12.
- It is the least common multiple of two one-digit even numbers.
- It is not a factor of 36.

A. 12   B. 24
C. 36   D. 48

2. Which of the following is NOT a correct statement?

A. If a is a multiple of b, then b is a factor of a.
B. If b is a factor of a, then b is a multiple of a.
C. Any two numbers can have a common multiple.
D. If a is a multiple of b and b is a multiple of c, then a is a multiple of c.

3. 48 is NOT a multiple of 36 because

A. 12 is the greatest common factor of 48 and 36.
B. 9 is a factor of 36 but not a factor of 48.
C. 48 is not a prime number.
D. No whole number multiplied by 36 will give a product of 48.

4. 18 is NOT a factor of 84 because

A. 84 is not prime.
B. 18 is not prime.
C. 6 is not a factor of 84.
D. 9 is not a factor of 84.

5. What is the smallest three-digit number divisible by 3?

6. If 2, 3, and 5 are factors of a number, list three other factors of the number.

7. Lisa believes that a characteristic of a prime number is that prime numbers are odd. Explain why Lisa's generalization is incorrect.

8. If 2 is not a factor of a number, why can't 6 be a factor of the same number?

9. One rectangle has an area of 48 cm^2. Another rectangle has an area of 80 cm^2. The dimensions of each rectangle are whole numbers. If each rectangle is to have the same length, what is the greatest possible dimension, in centimeters, the length can be?

10. Arlene believes that the larger a number is, the more factors the number has. Write an argument that agrees or disagrees with Arlene's belief.

11. Chiang believes that the LCM of two numbers is always greater than either number. Write an argument that agrees or disagrees with Chiang's belief.

12. Stan and John begin a race at the same time. John runs a lap of the track in 8 minutes. Stan runs the same lap in 6 minutes. When is the first time that John and Stan will complete a lap together?

1 B 5 Order of Operations

When simplifying an expression that involves more than one operation, use the following **order of operations**:

- Perform all operations within parentheses, and above and below a fraction bar.
- Evaluate all exponents and roots.
- Multiply or divide in order from left to right.
- Add or subtract in order from left to right.

 Model Problem

Evaluate $12 + 6 \div 3 - (5 - 2)$.

A. 12
B. 11
C. 7
D. 3

Solution Perform operations within parentheses first.

$$12 + 6 \div 3 - 3$$

Multiply and divide in order from left to right.

$$12 + 2 - 3$$

Add and subtract in order from left to right.

$$14 - 3 = 11$$

Answer B

 **PRACTICE**

1. $3(12 - 5) - 2 \times 6 =$

A. 9   B. 19
C. 48   D. 114

2. $84 - 18 \div (3^2 \times 2) =$

A. 3.3   B. 22
C. 78   D. 83

3. Evaluate the algebraic expression $2a - 3b$ if $a = 8$ and $b = 3$.

   A. 7   B. 10
   C. 30   D. 39

4. Which expression represents the following situation? You have six pages of postage stamps, each page has twelve stamps, and five stamps are missing from one page.

   A. $6 + 12 - 5$ B. $6 \times 12 - 5$
   C. $6(12 - 5)$ D. $6 \times 12 + 5$

5. Which expression has value 0?

   A. $4 \times (9 - 9)$ B. $4 \times 9 - 9$
   C. $(4 \times 9) - 9$ D. $9 - 9 \times 4$

6. An adult ticket to the planetarium costs \$9 and a child's ticket costs \$5. Write an expression to find the cost for 4 adult tickets and 3 children tickets. (Do not solve the expression.)

7. Insert parentheses to make the sentence true.

$$17 + 3 \div 4 + 1 = 4$$

8. Bob thinks the answer to the following problem is 1.

$$6^2 \div 3 - 9 \div 3$$

Sue disagrees. She thinks the answer should be 9. With whom do you agree? Explain.

Assessment Macro B

1. $9 + 12 \div 3 \times 5^2 - 19$

   A. 30   B. 51
   C. 90   D. 156

2. Which number below is not equivalent to 0.6?

   A. $\dfrac{3}{5}$

   B. $\sqrt{0.36}$

   C. 6%

   D. 6×10^{-1}

3. The distance on the number line between a number x and -5 is 7 units. Find all possible values for x.

   A. $-12, 2$ B. $12, -2$
   C. -12 D. -2

4. Two whole numbers are called **relatively prime** if their only common factor is 1. For example, 5 and 8 are relatively prime. How many of the following pairs represent numbers that are relatively prime?

   8 and 9, 18 and 21, 60 and 99, 51 and 200

   A. 1   B. 2
   C. 3   D. 4

5. Which of the numbers below has all of the following characteristics?

- It is a multiple of 9.
- It is a factor of 144.
- It is a perfect square.

   A. 18   B. 36
   C. 49   D. 72

6. Select the correct comparison between $(0.9)^2$ and $(0.9)^3$.

   A. $(0.9)^2 < (0.9)^3$
   B. $(0.9)^2 > (0.9)^3$
   C. $(0.9)^2 = (0.9)^3$
   D. The comparison cannot be determined without more information.

7. Nicole is preparing dinner. She needs 30 minutes to prepare a recipe. The dish needs to cool for $\frac{1}{4}$ hour before she can serve it. The recipe requires that the dish cook at 425° for 20 minutes and 375° for 20 minutes. If Nicole plans to serve dinner at 7:15, at what time should she begin to prepare the recipe?

A. 5:45   B. 5:50
C. 5:55   D. 6:05

8. For which of the following are the numbers NOT in correct order from SMALLEST to LARGEST?

A. $-3.1, -\frac{7}{2}, 0, 0.7, \frac{3}{2}$

B. $-2.3, -\frac{3}{2}, 0.1, \frac{1}{3}, 0.35$

C. $-6.1, -\frac{16}{3}, -4, -\frac{1}{4}, 0$

D. $-2.3, -\frac{3}{2}, -0.9, 0, 0.9$

9. Between which two numbers does $\frac{5}{9}$ fall?

A. $\frac{1}{2}$ and $\frac{11}{20}$ B. $\frac{1}{2}$ and $\frac{3}{5}$

C. $\frac{3}{5}$ and $\frac{3}{4}$ D. $\frac{14}{25}$ and $\frac{3}{5}$

10. What must be added to $(-3)^3$ to produce a sum of 0?

A. -27 B. 0
C. 9   D. 27

11. If $r = -2$ and $t = -5$, find the value of $r^4 + t$.

12. How many ratios of two whole numbers are equivalent to $\frac{3}{11}$ and have a two-digit denominator?

13. Calculate this quotient using scientific notation: $\dfrac{20,000 \times 300,000}{0.000005}$.
Express the answer in scientific notation. Show your procedure.

14. How many prime numbers are factors of 210?

15. How many two-digit numbers are multiples of 3 and also factors of 81?

16. A fraction is equivalent to $\frac{4}{5}$. The sum of the numerator and the denominator of the fraction is 36. What is the fraction?

17. The temperature at 5:00 A.M. was $-2°$ C. By 2:00 P.M. the temperature rose 13° and by 7:00 P.M. it had dropped 9°. What was the temperature at 7:00 P.M.?

A. 2°   B. 4°
C. 6°   D. 20°

18. The area of a square is 144 square feet. What is the perimeter of the square?

A. 12 feet   B. 24 feet
C. 48 feet   D. 72 feet

19. Write two different algebraic expressions to find the area of the given rectangle.

20. Explain why a multiple of 17, greater than 17, cannot be a prime number.

21. Leonardi is making up fruit baskets from 45 apples and 54 pears. Each basket must contain the same combination of apples and pears as all of the other baskets. What is the greatest number of baskets that he can make up using all the fruit? Explain how you arrived at your answer.

22. If a number is divisible by 8, is twice the number divisible by 8? Explain.

23. Given the four-digit number 8,4△3, can you find a value for △ such that the number will be divisible by 5? Explain your response.

24. Two nearby lights are flashing, one every 7 seconds and the other every 5 seconds. How many seconds long are the intervals between simultaneous flashings? Explain how you arrived at your answer.

25. Plastic spoons come in packages of 20, plastic forks in packages of 15, and plastic knives in packages of 12. A place setting requires a knife, a spoon, and a fork. What is the least number of packages of spoons, forks, and knives you can buy in order to get an equal number of each? Explain your work.

Open-Ended Questions

---

26. Find two sets of three numbers, each of which satisfies all these clues:

- Each number is a two-digit number.
- The greatest common factor for the three numbers is 6.
- The largest number is twice the smallest number.

27. A palindrome is a number that reads the same from left to right and from right to left. For example, 737 and 8,228 are palindromes.

a. How many palindromes are there between 2,000 and 5,000? Explain an efficient procedure for arriving at the answer.
b. What fraction of these palindromes are multiples of 3?
c. What is the first palindrome greater than 10,000 that also is a multiple of 3?

28. Examine the twenty-four fractions shown.

$\frac{1}{2}$	$\frac{1}{3}$	$\frac{1}{4}$	$\frac{1}{5}$	$\frac{1}{6}$	$\frac{1}{7}$	$\frac{1}{8}$	$\frac{1}{9}$
$\frac{2}{3}$	$\frac{2}{5}$	$\frac{2}{7}$	$\frac{2}{8}$	$\frac{2}{9}$	$\frac{2}{10}$	$\frac{2}{11}$	$\frac{2}{12}$
$\frac{3}{4}$	$\frac{3}{5}$	$\frac{3}{6}$	$\frac{3}{7}$	$\frac{3}{8}$	$\frac{3}{9}$	$\frac{3}{10}$	$\frac{3}{11}$

- Eliminate all fractions that are not in simplest form.
- Eliminate all fractions with values that are less than 0.4.

a. How many fractions remain?
b. Write the remaining fractions in order from least to greatest.
c. How do you know that no two of the remaining fractions can be equivalent to one another?
d. How do you know that no three of the remaining fractions can have a sum less than 1?

Apply ratios, proportions, and percents in a variety of situations.

1 C 1 Ratio and Proportion

A **ratio** is a comparison of two numbers by division. The ratio of two numbers a and b (where $b \neq 0$) can be expressed as:

$$a \text{ to } b \qquad \text{or} \qquad a{:}b \qquad \text{or} \qquad \frac{a}{b}.$$

A ratio that compares two unlike quantities, such as miles and hours, is called a **rate**.

Examples

  a. The ratio of 3 inches to 7 inches is 3:7.

  b. The ratio of 3 inches to 1 foot is 3:12 or 1:4.

  c. The ratio of diameter to radius is 2:1.

  d. The ratio of shaded squares to unshaded squares is 6:6 or 1:1.

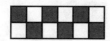

  e. The ratio of 55 miles to 1 hour is the rate 55 miles per hour.

A **proportion** is a statement that two ratios are equal. A proportion shows that the numbers in two different ratios compare to each other in the same way.

The proportion $\frac{2}{3} = \frac{8}{12}$ is read *2 is to 3 as 8 is to 12.*

In a proportion, the **cross products** are equal. In the proportion above, 3 and 8 are the **means**, or the terms in the middle of the proportion. 2 and 12 are the **extremes**, or the terms at the beginning and end of the proportion. In a proportion,

    the product of the means = the product of the extremes.

$$3 \times 8 = 2 \times 12$$
$$24 = 24$$

Example If pencils cost $2 for 6, how many pencils can you buy for $100?

Number of Pencils	6	x
Cost	$2	$100

The table can be used to write the proportion needed to solve the problem:

$$\frac{6}{2} = \frac{x}{100}$$

The proportion can be solved by setting cross products equal:

$$2x = 6 \cdot 100$$
$$2x = 600$$
$$x = 300 \text{ pencils}$$

Another approach is to extend the table, using numbers that are easy to calculate:

Number of Pencils	6	6(25) = 150	150(2) = 300
Cost	$2	$2(25) = $50	$50(2) = $100

> **To Find a Unit Rate:**
> 1. Set up a ratio comparing the given units.
> 2. Divide to find the rate for one unit of the given quantity.

Model Problem

1. Find the unit rate if you travel 150 miles in 2.5 hours.

   Solution Set up the ratio and divide.

   $$\frac{\text{miles} \rightarrow}{\text{hours} \rightarrow} \frac{150}{2.5} = \frac{1,500}{25} = \frac{60}{1}$$

   Note: In this case, multiplying the ratio by $\frac{10}{10}$ makes the division easier.

   Answer The rate is 60 mph.

2. A store has a 10 oz package of oat cereal for $2.29 and a 15 oz package of the same cereal for $2.89. Which is the better buy?

   Solution Find the price per one ounce of each package and compare.

   $$\frac{\text{price} \rightarrow}{\text{oz} \rightarrow} \frac{2.29}{10} = \frac{0.229}{1} = \$0.229/\text{oz}$$

   $$\frac{\text{price} \rightarrow}{\text{oz} \rightarrow} \frac{2.89}{15} = \frac{0.193}{1} = \$0.193/\text{oz}$$

   Answer The 10 oz package costs about 23 cents per ounce. The 15 oz package costs about 19 cents per ounce. The 15 oz package is the better buy.

3. Solve the proportion: $\frac{1.2}{1.5} = \frac{x}{5}$.

Solution Set the cross products equal and solve the equation for x.

$$(1.5)(x) = (1.2)(5)$$

$$1.5x = 6$$

$$\frac{1.5x}{1.5} = \frac{6}{1.5}$$

$$x = \frac{6}{1.5}$$

Answer $x = 4$

Note: A table may also be used to solve the proportion as follows:

1.2	1.2(10) = 12	12 ÷ 3 = 4	4
1.5	1.5(10) = 15	15 ÷ 3 = 5	5

4. In the scale on a map, 1 cm represents an actual distance of 250 km. What is the actual distance represented by a length of 1.75 cm?

Solution Set up the proportion and solve.

$$\frac{\text{cm} \rightarrow}{\text{km} \rightarrow}\frac{1}{250} = \frac{1.75}{x}$$

$$x = (1.75)(250)$$

Answer $x = 437.5$ km

PRACTICE

1. In a class of 25 students, there are 13 boys. What is the ratio of girls to boys?

A. 12:13
B. 12:25
C. 13:12
D. 13:25

2. Eric, a delivery driver, is paid at the rate of $8.50 an hour for the first 40 hours a week that he works. He is paid time and a half for any hours over 40. What would Eric's gross pay be for a week in which he worked 48 hours?

A. $340
B. $408
C. $442
D. $610

3. A basketball player makes 3 out of every 5 of her foul shots. At this rate, if she attempts 55 foul shots, how many will she miss?

A. 50
B. 40
C. 33
D. 22

4. In Beth's secret recipe, 4 eggs are needed to make 36 cranberry muffins. How many eggs are needed to make 90 muffins?

A. 8
B. 9
C. 10
D. 12

5. If 30 cards can be printed in 40 minutes, how many hours will it take to print 540 cards at the same rate?

A. 6 hours
B. 9 hours
C. 12 hours
D. 15 hours

6. Which proportion does NOT represent the given question?
If 48 oz cost $1.89, what will 72 oz cost?

A. $\frac{48}{1.89} = \frac{72}{x}$
B. $\frac{48}{72} = \frac{1.89}{x}$
C. $\frac{1.89}{72} = \frac{x}{48}$
D. $\frac{1.89}{48} = \frac{x}{72}$

7. With which roll of film would the cost of a single exposure be less? By how much less would it be?

> a roll of 20-exposure film for $2.30
> a roll of 12-exposure film for $1.50

   A. 20 exposures, $0.08
   B. 20 exposures, $0.01
   C. 12 exposures, $0.80
   D. 12 exposures, $0.01

8. Which of the following is a better buy? Explain why.

> a 3-pack of blank videotapes for $8.85
> a 2-pack of blank videotapes for $5.95

9. Samantha's job requires 40 hours of work per week. Which of the following situations represents a better annual salary offer? Explain why.

> a salary of $504.50 per week
> $12.50 per hour for 40 hours per week

10. Mario's car travels 468 miles on 18 gallons of gas. Will 40 gallons of gas be enough for Mario to travel 1,200 miles? Explain.

11. Solve the given proportion: $\dfrac{28}{32} = \dfrac{x}{40}$.

12. The scale on a map is $\dfrac{1}{2}$ inch = 55 miles. How far apart are two cities that are shown as being 5 inches apart on the map?

---

1 C 2 Percent

Percent (symbolized by %) means *for each hundred*. A percent is a ratio whose second term is 100. A percent can be written as a fraction or as a decimal.

$$25\% = \frac{25}{100} = 0.25 = \frac{1}{4}$$

In general, percent applications involve three terms:

percentage: part of the total   **rate**: percent   **base**: total amount

> To solve percent problems, use the formula: **percentage = rate × base**.

An alternative method for solving percent problems is to set up and solve the following proportion:

$$\frac{\text{Part}}{\text{Total}} = \frac{\text{Percent}}{100}$$

> **Key Percents to Remember**
> ---
> 100% is all.
> 50% is one-half.
> 25% is one-quarter.
> 10% is one-tenth.
> 1% is one-hundredth.
> 200% is double.

Types of Percent Problems

There are three basic types of percent problems. The following examples show both methods of solution.

1. Finding a percent of a number

What is 25% of 48?

Percentage = Rate × Base	$\dfrac{\text{Part}}{\text{Total}} = \dfrac{\text{Percent}}{100}$
Percentage = 0.25×48 Percentage = 12	$\dfrac{\text{Part}}{48} = \dfrac{25}{100}$ $\text{Part}(100) = (48)(25)$ $\text{Part}(100) = 1{,}200$ $\text{Part} = 12$

2. Finding a total if you know the percent and percentage

12 is 25% of what number?

Percentage = Rate × Base	$\dfrac{\text{Part}}{\text{Total}} = \dfrac{\text{Percent}}{100}$
$12 = 0.25 \times b$ $\dfrac{12}{0.25} = b$ $48 = b$	$\dfrac{12}{\text{Total}} = \dfrac{25}{100}$ $\text{Total}(25) = (12)(100)$ $\text{Total}(25) = 1{,}200$ $\text{Total} = 48$

3. Finding the percent when you know the part and total

12 is what percent of 48?

Percentage = Rate × Base	$\dfrac{\text{Part}}{\text{Total}} = \dfrac{\text{Percent}}{100}$
$12 = r \times 48$ $\dfrac{12}{48} = r$ $\dfrac{1}{4} = r$ $25\% = r$	$\dfrac{12}{48} = \dfrac{\text{Percent}}{100}$ $(12)(100) = 48(\text{Percent})$ $1{,}200 = 48(\text{Percent})$ $\dfrac{1{,}200}{48} = \text{Percent}$ $25 = \text{Percent}$

There are many real world applications of percent. Some of these applications are shown here.

Applications of Percent		
Type	**Problem**	**Solution**
Sales Tax	What is the total cost of a $30 purchase including a 5% sales tax?	Total = $30 + 0.05($30) = $30 + $1.50 = $31.50
Discount	A $20 shirt is on sale for 10% off. What is the sale price?	Sale Price = $20 − 0.1($20) = $20 − $2 = $18 = $18
Commission	A real estate agent makes a 5% commission on the sale of a house. What is the commission on a house that costs $250,000?	Commission = 0.05($250,000) = $12,500
Tip	Jill wants to leave a 15% tip on a dinner check of $62. How much is the tip?	Tip = 0.15($62) = $9.30
Percent Decrease	What is the percent of decrease of the price of a shirt that went from $30 to $25?	$\text{Percent Decrease} = \dfrac{\text{decrease}}{\text{original amount}}$ $= \dfrac{5}{30} = \dfrac{1}{6} = 0.167$ = 16.7% decrease
Percent Increase	The bear population went from 504 to 630. What is the percent of increase?	$\text{Percent Increase} = \dfrac{\text{increase}}{\text{original amount}}$ $= \dfrac{126}{504} = 0.25$ = 25% increase
Simple Interest	How much simple interest will Tom earn on $700 in one year if the bank pays 5% interest?	$\text{Simple Interest} = \text{principal} \times \dfrac{\text{interest}}{\text{rate}} \times \dfrac{\text{time}}{\text{(years)}}$ = $700 × 0.05 × 1 = $35

Model Problem

1. At a sale, Dan paid $59 for a sweater whose price had been reduced 20%. What was the original price?

Solution

20% reduction → sale price is 80% of original price

$59 is 80% of original price

$$p = r \times b$$
$$59 = 0.80 \times b$$
$$\frac{59}{0.80} = b$$
$$\$73.75 = b$$

Answer   The original price was $73.75.

2. The price of a shirt increased from $20 to $25. Find the percent increase.

Solution

$$\text{percent increase} = \frac{\text{increase}}{\text{original amount}}$$

Expressed as a percent:

$$\text{percent increase} = \frac{25 - 20}{20}$$
$$= \frac{5}{20} = \frac{1}{4} = 25\%$$

Answer   The percent increase is 25%.

PRACTICE

1. The number of students in the eighth grade is 240. The figure is 112% of what it was the previous year. This means that:

A. 12 more students are in the eighth grade now.
B. The number of students in the eighth grade decreased from last year to this year.
C. The total population of the school increased.
D. The number of eighth graders increased from last year to this year.

2. Which of the following is NOT a correct statement?

A. 63% of 63 is less than 63.
B. 115% of 63 is more than 63.
C. $\frac{1}{3}\%$ of 63 is the same as $\frac{1}{3}$ of 63.
D. 100% of 63 is equal to 63.

3. On a math test of 25 questions, Miriam scored 72%. How many questions did Miriam get wrong?

A. 7
B. 9
C. 18
D. 20

4. Laura bought a softball glove at 40% off the original price. This discount saved her $14.40. What was the original price of the glove?

   A. $5.76   B. $20.16
   C. $36   D. $50.40

5. The sale price of a reclining chair is $510 after a 15% discount has been given. Find the original price.

   A. $76.50   B. $433.50
   C. $586.50   D. $600

6. The regular price of an exercise bike is $125. If it is on sale for 40% off the regular price, what is the sale price?

   A. $90   B. $80
   C. $75   D. $48

7. The Sound System Store is selling portable CD players for $\frac{1}{3}$ off the regular price of $63. Murphy's Discount Store is selling the same CD players at 25% off the regular price of $60. What is the lower sale price for the radio?

   A. $35   B. $42
   C. $45   D. $48

8. If a price is doubled, it is increased by ___%.

9. If you needed to compute $33\frac{1}{3}$% of $45 without a calculator, would it be better to represent $33\frac{1}{3}$% as a fraction or a decimal? Explain your response.

10. When the sales-tax rate decreased from 7% to 6%, how much less in sales tax did you pay in purchasing an item priced at $480?

11. To obtain a score of 75% on a test containing 40 questions, how many questions must Juan get correct?

12. On a 10 × 10 grid, 31 squares are shaded. How many squares would have to be shaded on a 5 × 10 grid in order to have the same percent of the total squares shaded?

13. The cost of a first-class stamp increased from 32¢ to 37¢. Find the percent of increase.

14. The enrollment in a school went from 750 students to 600 students over a 10-year period. Find the percent of decrease.

15. Using the simple interest formula, $I = PRT$, what is the interest on $400 borrowed at 6% for 2 years? (*Note*: I = interest, P = principal, R = interest rate, T = time in years.)

16. Norma receives a commission of 5% on her snowboard sales. Her gross sales for November were $12,500. How much more will she receive if her commission is raised to 7%?

17. Is there a difference between buying an item at 50% off the regular price versus buying an item reduced by 30%, and then reduced by an additional 20%? Give examples to justify your answer.

Assessment Macro C

1. Find the value of x.

$$\frac{9}{x} = \frac{63}{112}$$

   A. 7   B. 9
   C. 16   D. 58

2. Every 4 days, 180 cars come through the assembly line. At this rate, how many cars come through in 7 days?

   A. 45   B. 315
   C. 720   D. 1,260

3. Which of the following is a correct representation for the ratio of 12 ounces to 2 pounds?

   A. 6:1   B. 3:8
   C. 1:2   D. 8:3

4. A denim jacket was on sale at a 25% discount. This discount was worth a savings of $5.00. What was the original price of the jacket?

   A. $30   B. $25
   C. $20   D. $12.50

5. Which of the following does NOT represent a 50% increase?

   A. $3 → $4.50
   B. $120 → $180
   C. $9 → $18
   D. $6.50 → $9.75

6. A box of machine parts contains 3 times as many good parts as defective parts. There were exactly 72 parts in the box. How many of the parts were defective?

   A. 18   B. 24
   C. 48   D. 54

7. About 15 percent of the students were absent from school on Wednesday. If 595 students were in school on Wednesday, how many students are enrolled in the school all together?

   A. 89   B. 506
   C. 700   D. 3,967

8. The school board estimates that there will be a 2.5% increase in enrollment for next year. If the present enrollment is 8,657, what is the estimated enrollment?

   A. 10,822   B. 8,874
   C. 8,682   D. 8,660

9. Wally's Used Car Lot is selling a used van at 20% off the usual price of $4,770. Mr. L's Used Car Lot is selling the same van at $\frac{1}{3}$ off the usual price of $5,643. Which offers the lower price and by how much?

   A. Wally's by $54
   B. Wally's by $927
   C. Mr. L's by $54
   D. Mr. L's by $927

10. Surprise.com sells wrapping paper for $3.75 a roll. Find the cost of an order of 8 rolls including tax and shipping. Tax is charged at the rate of 6% and is charged only on the purchase, not the shipping.

Surprise.com Shipping Charges	
Cost of Paper (with tax)	Add
less than $10.00	$0.55
$10.01–20.00	$1.00
$20.01–30.00	$1.75
over $30.00	$2.25

A. $33.55 B. $33.66
C. $34.05 D. $36.46

11. Franklin is buying a computer system from a mail-order house. The sale price is $2,599. He must pay 6% sales tax and a 5.5% shipping and handling charge. (The shipping charge does not apply to the sales tax.) How much is his total cost for the computer with tax and shipping?

A. $298.89
B. $2,741.95
C. $2,754.94
D. $2,897.89

12. Melissa answered 15 questions correctly on a 45-question multiple-choice test. How many questions will she answer correctly on a 75-question multiple-choice test if she gets the same ratio of questions correct on it as she did on the shorter test?

A. 15 B. 25
C. 45 D. 60

13. A photograph has a ratio of length to width of 8 to 5. If the width is 35 cm, how many centimeters is the length of the picture?

14. Judy bought a digital camera for $410. If sales tax was 7%, what was the total purchase price?

15. The cost of a two-scoop ice cream cone went from $3.50 to $3.85. Find the percent of increase.

16. Using the simple interest formula, $I = PRT$, what is the interest on $350 borrowed at 5% for 3 years?

17. The scale on a map is $\frac{1}{2}$ inch = 65 miles. How far apart are two cities that are $6\frac{1}{2}$ inches apart on the map? Explain your answer and approach.

Open-Ended Questions

18. The price of a certain brand of cough drops was increased as follows:

Before:	18 cough drops for $0.50
After:	16 cough drops for $0.60

Reggie claims that the percent increase is 20%. He figured this out by taking the difference of $0.10 and determining that $0.10 is $\frac{1}{5}$ or 20% of the original $0.50. Wendy claims that Reggie is wrong because he did not take into account the difference in the number of cough drops. Determine the correct percent increase. Explain your procedure.

19. Markdown City offers successive discounts of 10%, 15%, and 20% on a purchase. These discounts can be applied in any order. If you were the customer, which would be the best order for the three discounts to be applied? Defend your response.

20. The formula for the volume of a rectangular solid is

$$V = lwh.$$

a. What happens to the volume of a rectangular solid if the length is increased by 10%, the width is increased by 10%, and the height is also increased by 10%?

b. In changing the dimensions of a rectangular solid, the length and width are each increased by 10%. The height is decreased by 20%. Brigit says that the result is that the volume remains the same. Stella says the volume is decreased by a very slight percent. Who is correct? Explain.

Assessment Cluster 1

1. Which set of numbers is in order from LEAST to GREATEST?

A. $\frac{7}{8}$ 0.6 $\frac{3}{4}$ $\frac{4}{5}$

B. $\frac{4}{5}$ $\frac{7}{8}$ $\frac{3}{4}$ 0.6

C. $\frac{7}{8}$ $\frac{4}{5}$ $\frac{3}{4}$ 0.6

D. 0.6 $\frac{3}{4}$ $\frac{4}{5}$ $\frac{7}{8}$

2. Ted works 40.5 hours per week as a golf caddie. He is paid $6.50 per hour. What is Ted's gross pay for the year?

A. $263.25   B. $3,159
C. $6,318   D. $13,689

3. Which property did Joe use in figuring out the problem?

$$9 \times 16 + 9 \times 14 = 9(16 + 14)$$
$$= 9(30) = 270$$

A. associative property of addition
B. associative property of multiplication
C. distribute property
D. commutative property of addition

4. Which of the following is NOT equal to the other three?

A. 1.5×10^1 B. $\frac{15}{10}$

C. 150%   D. $\sqrt{2.25}$

5. Which of the following numbers is between $\frac{1}{10,000}$ and $\frac{1}{100,000}$, and correctly expressed in scientific notation?

A. 4.5×10^{-3}
B. 4.5×10^3
C. 4.5×10^{-4}
D. 4.5×10^{-5}

6. A $32 sweater is reduced by 25% for a holiday sale. By what percent must the sale price of the sweater be multiplied to restore the price to the original price before the sale?

A. $133\frac{1}{3}\%$ B. 125%
C. 25%   D. 8%

7. The members of the Decorating Committee for the Valentine's Day dance have 24 red carnations, 32 white carnations, and 40 pink carnations. They want to form as many identical centerpieces as possible, using all of the carnations so that each centerpiece has the same combination of colors as all of the other centerpieces. How many pink carnations should each centerpiece have?

A. 4   B. 5
C. 8   D. 10

8. Which of the following is NOT a way to find 120% of a number?

   A. Multiply the number by 1.20.
   B. Divide the number by 5 and add the result to the number.
   C. Divide the number by 5 and multiply the result by 6.
   D. Multiply the number by 0.20 and multiply the result by 5.

9. On some days, a bakery packages cupcakes 4 to a box. On other days, they package cupcakes 6 or 8 to a box. On a given day, all of the cupcakes baked were packaged and there was one cupcake left over. Which of the following could NOT be the number of cupcakes baked on that day?

   A. 22   B. 25
   C. 49   D. 97

10. There are three times as many girls as boys in the Spanish Club of Liberty Middle School. If there are 36 members in the club, how many of them are boys?

   A. 9   B. 12
   C. 15   D. 27

11. $12 \div (9 - 7)^2 \times 8 + 24 \div 6 =$

   A. 292   B. 28
   C. 16   D. 8

12. Mr. Kim, a furniture salesperson, is paid $300 a week plus commission. His commission is 5% of his weekly sales. His weekly sales for the month of January are shown in the table.

Week 1	Week 2	Week 3	Week 4
$8,576	$9,500	$7,362	$10,567

   What was his average weekly commission in January?

   A. $429   B. $450
   C. $528   D. $1,800

13. We know that $2^3 = 2 \times 2 \times 2 = 8$. What number in the box would make the following TRUE?

   $$8^6 = 2^{\square}$$

   A. 9   B. 15
   C. 18   D. 24

14. Miranda earns $7.50 an hour for the first 40 hours a week she works. She earns time and a half for any hours over 40 she works during the week and double time for hours worked on the weekend. Her time card for one week is shown below. How much did Miranda earn?

Day	Hours
Monday	$8\frac{1}{2}$
Tuesday	9
Wednesday	$9\frac{1}{2}$
Thursday	7
Friday	$3\frac{1}{2}$

   A. $341.25   B. $361
   C. $375   D. $382.50

15. A fraction is equivalent to $\frac{3}{8}$. The sum of the numerator and denominator is 33. What is the fraction?

16. Steven visited Canada and took $325 to spend. The rate of currency exchange was $1.1515 Canadian dollars per U.S. dollar. To the nearest dollar, how many Canadian dollars did Steven get for the exchange?

17. Helene is paid at the rate of $6.50 an hour for the first 40 hours a week that she works. She is paid time and a half for any hours over 40. How much more will Helene make working 50 hours compared with working 46 hours?

18. Daniel inherited a collection of baseball cards that has 80 all-star cards and 120 rookie cards. His mother told him to give part of the collection to Robbie, his younger brother. She said that Daniel did not have to give away half the collection, but had to give his brother the same ratio of all-stars to rookies as he kept for himself. Daniel decided to give Robbie 20 all-star cards. How many rookie cards did he have to give Robbie?

19. A baseball player's batting average is the ratio of the number of hits to the official number of times at bat. A player had 150 hits during a season and wound up with a batting average of .300. What was the total number of times he was at bat for the season?

20. Rachel lost her cat and wants to put up posters to find it. The copy store charges 4.2 cents per copy. There is an additional charge of one-half cent for each copy on colored paper. How much will it cost Rachel to make 100 copies on white paper and 100 copies on blue paper?

21. Cherry tomatoes sell for $2.19 per pound. If a bag of cherry tomatoes costs $4.25, what is the weight of the bag of tomatoes to the nearest hundredth of a pound?

22. Because of a printer malfunction, Harold could not read some of the information on the monthly statement from his checking account. From the information in the Account Summary, find the ending balance of Harold's account.

Account Summary	
Account Number 00-537-387-5	
Beginning balance 08/01	875.45
3 deposits/credits	xxxxxx
3 checks/debits	xxxxxx
Service fee	xxxxxx
Ending balance 08/31	xxxxxx
Transactions	
Date	**Amount**
08/02	+135.00
08/05	−163.50
08/11	−14.75
08/22	−35.92
08/25	+214.35
08/31 Service fee	−11.45
08/31 Interest	+3.45

23. A discount of 20% followed by a discount of 15% is equivalent to what single discount? Use an example to demonstrate your answer.

24. A recipe calls for 2 quarts of juice to 3 quarts of seltzer to make a punch. How much juice would be needed to make 8 gallons of punch?

25. The cost of a cheeseburger increased from $2.00 to $2.25. Determine the percent of increase. Explain your approach.

26. In Hope City, the sales tax rate increased from 7% to 7.25%. Under this change, how much additional tax would a shopper in Hope City pay when purchasing an item priced at $800? Show your process in determining the answer.

27. The scale on a map is $\frac{1}{2}$ inch = 80 miles. What is the distance between two cities that are $4\frac{3}{4}$ inches apart on the map? Show how you arrived at your answer.

28. If 4, 5, and 7 are factors of a number, list four other numbers that would also be factors of the number.

Open-Ended Questions

---

29. Your friend applies the distributive property to multiplication and determines $2(3 \times 5) = 2(3) \times 2(5)$. Write a paragraph explaining how you would convince your friend that he is incorrect.

30. Every Monday, at Capri Pizza, lucky customers can get free slices of pizza and free soda. Every 10th customer gets a free plain slice of pizza and every 12th customer gets a free cup of soda. On the first Monday of March, Capri Pizza had 211 customers.

   a. How many free pizza slices were given away? How many cups of soda were given away?

   b. When Jackie walked into Capri Pizza, the owner told her that she was the first person to get both free items (the free slice of pizza and the free soda). What number customer must Jackie have been? How many of the other 211 customers would also get both free items?

   c. Suppose that the owner also decides to give away a free bag of potato chips to every 7th customer. Would any of the 211 customers be lucky enough to get all three free items? Explain.

31. Your local supermarket offers two brands of cheese sticks:

> Brand A: 12-ounce package for $2.49
> Brand B: 15-ounce package for $3.19

   a. Which is the better buy? Show how you arrived at your answer.

   b. Suppose the one that is a better buy now has a 10% price increase. Is it still the better buy? Explain.

32. A picture measures 5 inches by 7 inches. The picture is enlarged to a size of 15 inches by 21 inches.

   a. What are the area and perimeter of the original picture?

   b. What are the area and perimeter of the enlarged picture?

   c. What is the percent of increase of the area of the enlarged picture?

33. Use the number line to answer the questions that follow.

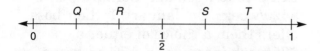

a. Write a fraction that could be a possible value for each of the four points Q, R, S, and T.

b. Write a decimal that could be a possible value for each of the four points Q, R, S, and T.

c. Name a possible decimal between points Q and R.

d. Name a possible fraction between points S and T.

Extra Practice

Open-Ended Questions

Study Questions 1 and 2, which are presented with their solutions, and then try Questions 3 and 4.

When responding to an open-ended question, think about what you must do to form a response that will receive a score of 3.

- Answer all parts of the question.
- Present your work clearly, so that the person grading it will understand your thinking.
- Show all your work, including calculations, diagrams, and written explanations.

1. In January, a store sells an item for $60.00. In February, over a four-week period, the store offers 10% price decreases on the price during the week before. For example, for the first week of February, the price was $60.00 minus 10% for a new price of $60.00 − $6.00 or $54.00.

- Find the new prices for the other three weeks in February.
- How does the new price for the fourth week compare with a single

price decrease of 40% off the original $60.00 price?

Solution (for a score of 3)

Week 1: $54.00

Week 2: $54.00 − 10% of $54.00
= $48.60

Week 3: $48.60 − 10% of $48.60
= $43.74

Week 4: $43.74 − 10% of $43.74
= $39.37

A single price decrease of 40% off the original price:

$$40\% \text{ of } \$60.00 = \$24.00$$
$$\$60.00 - \$24.00 = \$36.00$$

As a result, the new price is less than the final price after the four separate 10% decreases ($39.37). The single 40% decrease is better for the customer.

2. The costs of different brands of peanut butter are shown in the chart:

Brand A	12-ounce jar	$1.32
Brand B	18-ounce jar	$2.07
Brand C	32-ounce jar	$3.84

- Which is the best buy? Explain.
- If you have a recipe that calls for 36 ounces of peanut butter, describe the most economical way to purchase this amount. Explain your reasoning.
- If Brand A wants to offer a 15-ounce jar at the same per ounce price as the 12-ounce jar, what should be the cost of this 15-ounce jar?

Solution   (for a score of 3)

Using the amounts in the chart, find the cost per ounce for each brand by dividing the cost of the jar by the number of ounces.

Brand A	$0.11/oz
Brand B	$0.115/oz
Brand C	$0.12/oz

The best buy is Brand A.

Buying 3 of Brand A would give you 36 ounces at a cost of 3 × $1.32 = $3.96. Buying 2 of Brand B would also give you 36 ounces, but at a cost of 2 × $2.07 = $4.14. You would need two jars of Brand C, giving you 64 ounces. This would not be a good choice. Buying 3 of Brand A would be the most economical.

The cost of the 15-ounce jar should be $1.65.

$$\frac{12}{1.32} = \frac{15}{x}$$

$$12x = 15(1.32)$$

$$12x = 19.80$$

$$x = 1.65$$

3. Thomas and Bernard are the only two employees at a bicycle shop. Thomas earns $550/week and Bernard earns $420/week. On January 1, their boss offers them a choice of either a $35/week increase or a 7% increase.

- Which increase is the best choice for Thomas? Explain your thinking.
- Which increase is the best choice for Bernard? Explain your thinking.
- If their boss offers as a third choice a bonus of $1,500 for the year (52 weeks) instead of a pay increase, should Thomas take the bonus? Explain why or why not.

4. Alexander's Creamery is selling a new flavor of banana chip ice cream. The ice cream costs $2.80 per quart to manufacture and is packaged in small, medium, and large containers at the following prices:

Small	one cup for $1.50
Medium	one pint for $3.96
Large	one quart for $4.72

- For a consumer, which size is the best buy? Show how you got your answer.
- Which size is the most profitable for the Creamery? (Use the formula Profit = Selling Price – Manufacturing Costs.) Show your work.
- After a month of poor sales, the Creamery decides to lower the price of the small one-cup packages. What is the lowest price they can charge and still make 5 cents profit per container?

Spatial Sense and Geometry

Recognize, identify, and represent spatial relationships and geometric properties.

2 A 1 Geometric Terms

Term	Illustration	Definition
point	•A	a basic undefined term, a location in space (A point has no dimensions.)
line ($\overleftrightarrow{AB}$)	⟵•——————•⟶ A B	a basic undefined term, extends infinitely in two directions (A line has one dimension.)
ray ($\overrightarrow{AB}$)	•——————•⟶ A B	the part of line AB that contains point A and all the points of AB that are on the same side of point A as point B

Term	Illustration	Definition
line segment ($\overline{AB}$)	A ●————● B	two points and all the points between them that lie on the line containing the two points
midpoint	A ●——M●——● B	the point on a line segment that divides it into two equal lengths
plane		a basic undefined term, a flat surface that extends infinitely (A plane has two dimensions.)
polygon		a closed plane figure formed by three or more line segments with common endpoints (Each side intersects exactly two other sides, but only at their endpoints.)
vertex of a polygon	A	a point where two adjacent sides meet
diagonal of a polygon		a segment joining nonconsecutive vertices of the polygon
polyhedron		a closed 3-dimensional figure made up of flat polygonal regions
vertex	face / vertex	in a polyhedron, where 3 or more edges intersect
edge		in a polyhedron, a line segment in which a pair of faces intersect
face	edge	in a polyhedron, the flat polygonal surfaces that intersect to form the edges of the polyhedron

Term	Illustration	Definition
angle ($\angle ABC$)		two rays that share a common endpoint
acute angle		an angle that measures between 0° and 90°
obtuse angle		an angle that measures between 90° and 180°
right angle		an angle that measures 90°
straight angle		an angle that measures 180°
angle bisector		a ray that divides an angle into two congruent angles
adjacent angles		two angles that have the same vertex and a common side between them $\angle 1$ and $\angle 2$ are adjacent. $\angle 2$ and $\angle 3$ are adjacent. $\angle 1$ and $\angle 3$ are not adjacent.
vertical angles		a pair of congruent angles, formed by two intersecting lines, that have a common vertex and are not adjacent $\angle 1 \cong \angle 3$ $\angle 2 \cong \angle 4$

Term	Illustration	Definition
complementary angles		two angles whose measures have a sum of 90°
supplementary angles	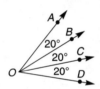	two angles whose measures have a sum of 180°

Model Problem

1. Which of the following statements is true about the diagram?

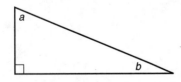

A. $\overrightarrow{OB}$ is the only angle bisector.
B. $\overrightarrow{OC}$ is the only angle bisector.
C. $\overrightarrow{OB}$ and $\overrightarrow{OC}$ are angle bisectors.
D. There are no angle bisectors.

Solution Since m$\angle AOC$ = 40 and m$\angle AOB$ = m$\angle BOC$ = 20, then $\overrightarrow{OB}$ is an angle bisector. Similarly, $\overrightarrow{OC}$ bisects $\angle BOD$.

Answer C

2. Explain why $\angle a$ is the complement of $\angle b$.

Solution Since the sum of the measures of the three angles of a triangle is 180° and the triangle contains a 90° angle, the sum of the other two angles must be 90°. Therefore, the two angles are complementary.

3. The measures of two supplementary angles are in the ratio of 2:3. What is the degree measure of the larger angle?

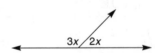

Solution Since supplementary angles have a sum of 180°, the relationship between the two angles can be expressed by the equation

$$2x + 3x = 180°$$
$$5x = 180°$$
$$x = 36°$$

The larger angle is $3x$, so $3(36) = 108°$.

Answer 108°

PRACTICE

1. The supplement of ∠z is 120°. What is the complement of ∠z?

   A. 30°   B. 60°
   C. 120°   D. 250°

2. The measures of two supplementary angles are in the ratio of 4:5. What is the degree measure of the smaller angle?

   A. 20°   B. 80°
   C. 100°   D. 160°

3. In which of the following diagrams are ∠a and ∠b vertical angles?

   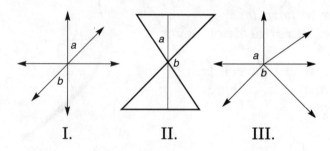

   I.   II.   III.

   A. I only
   B. I and II
   C. I and III
   D. I, II, and III

4. Which of the following statements is always TRUE?

   A. Vertical angles are supplementary.
   B. The complement of an obtuse angle is acute.
   C. The supplement of an obtuse angle is acute.
   D. Complementary angles are congruent.

5. Draw and label the figure described. Planes P and Q intersect each other. They both intersect plane R.

6. If D is the midpoint of $\overline{AC}$ and C is the midpoint of $\overline{AB}$, what is the length of $\overline{AB}$ if BD = 12 cm?

7. If $\overrightarrow{BD}$ is the angle bisector of ∠ABC and $\overrightarrow{BE}$ is the angle bisector of ∠ABD, what is m∠ABC if m∠DBE = 36?

   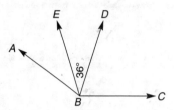

8. In the rectangular prism shown, how many vertices, edges, and faces are there?

   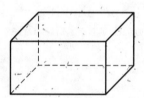

9. How many diagonals are there in a hexagon?

   A. 3   B. 6
   C. 9   D. 18

10. Find the measure of the angle made by the hands of a clock at 4:30. Classify the angle as acute, right, or obtuse.

11. $\overleftrightarrow{AB}$ and $\overleftrightarrow{BA}$ are two ways to name the line shown. Using two letters (from A, B, and C), how many total ways are there to name the line?

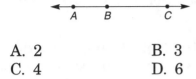

A. 2

B. 3

C. 4

D. 6

12. Explain why the following must be TRUE: Any time an obtuse angle is bisected, two acute angles are formed.

---

2 A 2 Geometric Relationships

Lines

In a plane, distinct lines will either *intersect* or be *parallel*. **Perpendicular lines** intersect at right angles. **Parallel lines** never intersect.

Lines will be parallel if certain relationships exist between pairs of angles formed when the lines are cut by a third line, called a **transversal**.

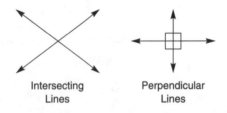

Intersecting Lines

Perpendicular Lines

Two Lines Are Parallel If:

1. alternate interior angles are congruent.

$\left.\begin{array}{l} \angle4 \text{ and } \angle6 \\ \angle3 \text{ and } \angle5 \end{array}\right\}$ alternate interior angles

2. corresponding angles are congruent.

$\left.\begin{array}{l} \angle1 \text{ and } \angle5 \\ \angle2 \text{ and } \angle6 \\ \angle4 \text{ and } \angle8 \\ \angle3 \text{ and } \angle7 \end{array}\right\}$ corresponding angles

3. interior angles on the same side of the transversal are supplementary.

$\left.\begin{array}{l} \angle4 \text{ and } \angle5 \\ \angle3 \text{ and } \angle6 \end{array}\right\}$ interior angles on the same side of the transversal.

Conversely, if you know the lines are parallel, then you know that the angle relationships are true.

Skew lines are lines that do not intersect and are not in the same plane.

Note: In geometric figures, parallel sides are sometimes represented by arrow symbols, while congruent sides are represented by tick marks.

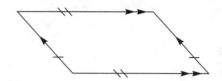

Model Problem

1. Line *a* is parallel to line *b*. Line *c* is perpendicular to line *b*. Line *d* is parallel to line *c*. Line *e* is perpendicular to line *d*. How is line *a* related to line *e*?

Solution   Draw a diagram to determine the relationship.

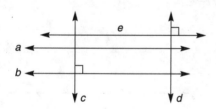

Answer   Line *a* is parallel to line *e*.

2. In the figure, ∠5 is congruent to ∠13 (in symbols, ∠5 ≅ ∠13). Based on that information, what pairs of lines can you conclude are parallel?

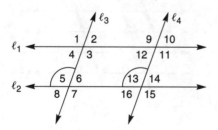

Solution   ∠5 and ∠13 are corresponding angles formed by l_3 and l_4 being cut by l_2. Since ∠5 ≅ ∠13, $l_3 \parallel l_4$. Since no additional information is provided about angles, you cannot conclude that l_1 is parallel to l_2.

3. If line *l* is parallel to line *m* and m∠2 = 40, find the measures of ∠5, ∠6, ∠7, and ∠8.

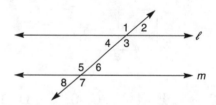

Solution   Since the lines are parallel and ∠2 and ∠6 are corresponding angles, the measure of ∠6 would also be 40. ∠5, the supplement of ∠6 would be 140°. ∠7, the supplement of ∠6, would also be 140°. Since ∠6 and ∠8 are vertical angles, m∠8 = m∠6 = 40.

PRACTICE

1. How many pairs of skew lines are represented by the edges of a cube?

   A. 0   B. 6   C. 12   D. 24

2. How many angles are formed when one transversal crosses four parallel lines?

   A. 8   B. 12   C. 16   D. 24

3. If ray *AB* is perpendicular to ray *AC*, what is the measure of ∠*EAG*?

   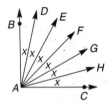

   A. 15°   B. 30°   C. 36°   D. 60°

4. *ACDF* is a rectangle divided into two squares. How many pairs of line segments in the figure are perpendicular?

   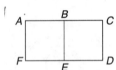

   A. 3   B. 4   C. 6   D. 14

5. Lines *l* and *m* are parallel. Transversal *t* is not perpendicular to line *l* or line *m*. Which of the following pairs of angles are NOT supplementary?

   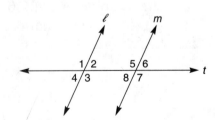

   A. ∠1 and ∠2   B. ∠2 and ∠5
   C. ∠5 and ∠7   D. ∠3 and ∠6

6. Based on the figure, which of the statements that follow is FALSE?

   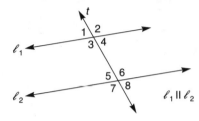

   A. ∠1 ≅ ∠4
   B. ∠1 ≅ ∠5
   C. ∠1 is supplementary to ∠3
   D. ∠1 ≅ ∠7

7. In the figure, if $l_1 \parallel l_2$, $l_2 \parallel l_3$, and $l_1 \perp l_4$, which of the following statements must be true?

   I. $l_1 \parallel l_3$

   II. $l_2 \perp l_4$

   III. $l_3 \perp l_4$

   A. I only   B. II only
   C. I and II   D. I, II, and III

8. How many pairs of parallel lines are shown?

2 A 3 Two-Dimensional Figures

Polygons

A **polygon** is a closed plane figure made up of line segments. Each side intersects exactly two other sides, but only at their endpoints. In a **regular polygon**, all sides and all angles are congruent.

Polygons can be classified by number of sides:

Number of Sides	Polygon
3	triangle
4	quadrilateral
5	pentagon
6	hexagon
7	heptagon
8	octagon
9	nonagon
10	decagon
n	n-gon

A **diagonal** is a line segment connecting any two nonadjacent vertices of a polygon. The sum of the angles of a polygon can be found by using diagonals to divide the polygon into triangles and then multiplying the number of triangles by 180°.

Example   A hexagon can be divided into 4 triangles, so the sum of the angle measures is 4(180°) = 720°.

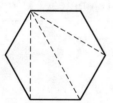

In general, the sum of the angle measures S of a polygon with n sides is given by the formula

$$S = 180°(n - 2).$$

ABCDE is a regular pentagon. What is the measure of each interior angle of the figure?

Solution   Since a pentagon has 5 sides, $n = 5$. Therefore, to find the sum of the measures of the interior angles of the pentagon, use the formula $S = 180°(n - 2)$.

$$S = 180°(5 - 2)$$
$$S = 180°(3) = 540°$$

Since the pentagon is regular, all angles are equal. Therefore, to find the measure of each interior angle, divide 540 by 5.

Answer   Each interior angle equals 108°.

Triangles

In a triangle, the sum of the lengths of any two sides must be greater than the length of the third side. The sum of the interior angles of a triangle is 180°.

Classification of Triangles			
By Number of Congruent Sides		**By Types of Angles**	
no sides congruent	*scalene triangle*	one right angle	*right triangle*
two sides congruent	*isosceles triangle*	one obtuse angle	*obtuse triangle*
all sides congruent	*equilateral triangle*	all acute angles	*acute triangle*

Examples

1. $\triangle LMN$ is an obtuse isosceles triangle.

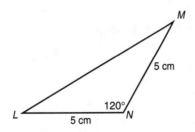

2. $\triangle QRS$ is a right scalene triangle.

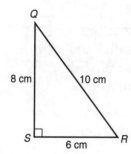

Q

8 cm   10 cm

S   6 cm   R

Model Problem

1. In $\triangle ABC$, m$\angle A = 52$ and m$\angle B = 12$. Classify $\triangle ABC$ as acute, right, or obtuse.

Solution The sum of the angles of the triangle is 180°.

$$52° + 12° = 64°$$
$$180° - 64° = 116°$$

m$\angle C = 116$. $\angle C$ is an obtuse angle.

Answer $\triangle ABC$ is an obtuse triangle.

2. Explain why a triangle cannot have sides with lengths of 4 cm, 7 cm, and 13 cm.

Solution In a triangle, the sum of the lengths of any two sides must be greater than the length of the third side.

13 cm + 7 cm > 4 cm
13 cm + 4 cm > 7 cm
7 cm + 4 cm is not greater than 13 cm.

Therefore, the triangle cannot have sides of 4 cm, 7 cm, and 13 cm.

Quadrilaterals

A **quadrilateral** is a four-sided polygon. Quadrilaterals have four interior angles, and the sum of these angles is always 360°. Special quadrilaterals are classified based on side lengths, angle measure, and parallel sides.

Special Quadrilaterals

Name	Illustration	Characteristics
parallelogram		quadrilateral with both pairs of opposite sides parallel and congruent; quadrilateral with both pairs of opposite angles congruent
rectangle		parallelogram with right angles
rhombus		parallelogram with all sides congruent
square		rectangle with all sides congruent
trapezoid		quadrilateral with one pair of opposite sides parallel
isosceles trapezoid		trapezoid with congruent legs

Model Problem

1. Which of the following statements is true?

   A. All rectangles are parallelograms.
   B. All parallelograms are rectangles.
   C. All quadrilaterals are trapezoids.
   D. All trapezoids are parallelograms.

Solution A rectangle is a special parallelogram with right angles. Therefore, any rectangle must be a parallelogram.

Answer A

2. A quadrilateral *ABCD* has vertices $A(0, 0)$, $B(6, 0)$, $C(7, 5)$, $D(1, 5)$. What is the best name for quadrilateral *ABCD*?

   A. rectangle
   B. trapezoid
   C. parallelogram
   D. rhombus

Solution Plot the points (all of which are in the first quadrant) and examine the properties of the resulting figure.

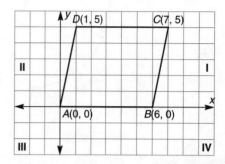

Since both pairs of opposite sides are parallel, the figure is a parallelogram. Since adjacent sides are not congruent, it is not a rhombus. Since there are no right angles, it is not a rectangle.

Answer C

Circles

A **circle** is a closed plane figure that represents all of the points a specified distance from a point called the **center**.

Key parts of the circle are pictured in the diagram.

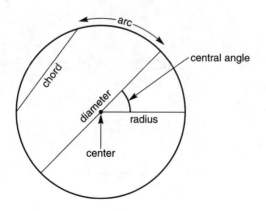

If two or more circles have the same center, they are known as **concentric circles**.

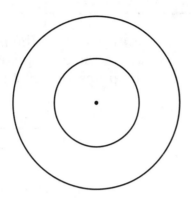

Model Problem

1. Knowing that a chord is a line segment joining two points on a circle, find the number of chords pictured in the figure.

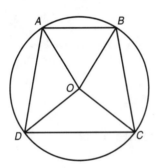

   A. 4   B. 6   C. 7   D. 8

Solution By the definition of a chord, $\overline{AB}$, $\overline{BC}$, $\overline{CD}$, and $\overline{AD}$ are chords. $\overline{AO}$, $\overline{BO}$, $\overline{CO}$, and $\overline{DO}$ are radii; they are not chords. Therefore, there are four chords pictured.

Answer A

2. A circle with a center at $(5, 0)$ passes through the point $(1, 0)$. What is the length of a diameter of the circle?

Solution Since the center is at $(5, 0)$ and $(1, 0)$ is on the circle, the length of a radius is the distance from $(5, 0)$ to $(1, 0)$, or 4 units.

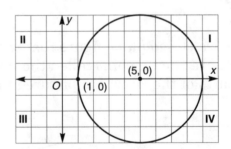

Answer A diameter has a length of 8 units.

1. Which of the following sets of lengths does NOT represent a triangle?

   A. 1, 1, 1   B. 3, 4, 5
   C. 4, 4, 8   D. 5, 12, 13

2. If the measure of one angle of a triangle is equal to the sum of the measures of the other two, then the triangle is always

   A. acute   B. obtuse
   C. right   D. isosceles

3. Brian claims he can make each of the following triangles.

   > isosceles right triangle
   > isosceles obtuse triangle
   > equilateral obtuse triangle
   > scalene right triangle

   Andre claims that only three are possible. Who is correct?

   A. Brian
   B. Andre
   C. both are incorrect
   D. cannot be determined

4. *ABCDEF* is a regular hexagon. The diagonals from vertex *A* are shown. How many isosceles obtuse triangles have been formed?

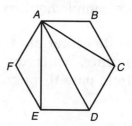

   A. 0
   B. 2
   C. 4
   D. cannot be determined

5. In parallelogram *ABCD*, what is the measure of ∠*C*?

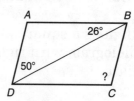

   A. 76°   B. 86°
   C. 100°   D. 104°

6. Quadrilateral *ABCD* has vertices at $A(-2, 2)$, $B(-2, -4)$, $C(5, -8)$, and $D(5, 6)$. What is the most specific name for the quadrilateral?

   A. parallelogram
   B. isosceles trapezoid
   C. trapezoid
   D. rectangle

7. Triangle *ABC* has vertices *A*(1, 6), *B*(10, 3), and *C*(1, 3). Which of the following must be TRUE concerning ∠*A* and ∠*B*?

   A. They are congruent.
   B. They are supplementary.
   C. One is twice as large as the other.
   D. They are complementary.

8. A circle has its center at (4, 0) and a radius of 5 units. Which quadrant(s) does the circle pass through?

   A. I only
   B. I and II
   C. I and IV
   D. all four quadrants

9. Which of the following statements are TRUE?

   I. A rhombus with right angles is a square.
   II. A rectangle is a square.
   III. A parallelogram with right angles is a square.

   A. I only
   B. II only
   C. III only
   D. I and III

10. A pentagon has two right angles. What is the sum of the measures of the other three angles?

   A. 108°   B. 360°
   C. 450°   D. 540°

11. Which of the following could NOT be the sum of the measures of the interior angles of a polygon?

   A. 720°   B. 1,080°
   C. 1,900°   D. 1,980°

12. A circle with its center at *P* has a radius of 5 cm. Line segment *PQ* measures 6 cm and line segment *PR*

measures 5.1 cm. Which of the following is a TRUE statement?

   A. Points *Q* and *R* are inside the circle.
   B. Points *Q* and *R* are on the circle.
   C. Points *Q* and *R* are outside the circle.
   D. The distance from point *R* to point *Q* is 0.9 cm.

13. Three concentric circles are shown. The diameter of the largest circle is 16 units. The diameter of the middle circle is 12 units and the diameter of the smallest circle is 10 units. What is the distance between the smallest circle and the middle circle?

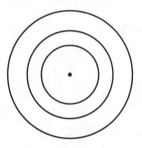

   A. 1   B. 2   C. 4   D. 6

14. A circular swimming pool has a diameter of 20 feet. Two poles for a volleyball net are each placed 15 feet from the center of the pool and as far apart as possible. What is the length of net needed to be strung tautly between the two poles? (Assume the net extends pole to pole.)

15. A trapezoid has three sides of length 5 inches. What two figures are formed by a diagonal of the trapezoid?

   A. two isosceles triangles
   B. two scalene triangles
   C. an isosceles triangle and a scalene triangle
   D. an isosceles triangle and a right triangle

16. Which of the following is NOT always true about the angles of an isosceles trapezoid?

A. The sum of the angle measures is 360°.

B. The figure has two acute and two obtuse angles.

C. Two of the angles are complementary.

D. The opposite angles are supplementary.

17. Explain why the measure of ∠a equals the measure of ∠c.

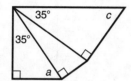

18. Starting with a rectangle *ABCD*, we can make the figure shown by removing side $\overline{AD}$ and drawing line segments $\overline{AE}$ and $\overline{ED}$ to form an obtuse angle.

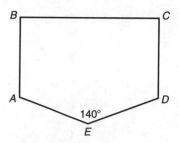

a. What type of polygon is *ABCDE*?

b. If the measure of ∠*AED* = 140°, what is the sum of the measures of ∠*EAB* and ∠*EDC*? Explain how you obtained your answer.

---

2 A 4 Three-Dimensional Figures and Spatial Relationships

Common Three-Dimensional Figures		
Name	**Illustration**	**Description**
pyramid		a polyhedron in which the base is a polygon and the lateral faces are triangles with a common vertex
prism		a solid with 2 faces (bases) formed by congruent polygons that lie in parallel planes and whose other faces are rectangles

To Visualize or Represent a Figure:

• Make a drawing—use graph paper or isometric dot paper.

• Think about a physical model.

• Create a physical model.

Name	Illustration	Description
cone		a solid consisting of a circular base and a curved lateral surface that extends from the base to a single point called the vertex
cylinder		a solid with congruent circular bases that lie in parallel planes
sphere		the set of all points at a given distance from a given point

Model Problem

1. One way of looking at a three-dimensional figure is to make a **net**, a flat pattern that can be used to construct the figure. Which of the following cardboard nets cannot be folded along the dotted lines to make a closed box?

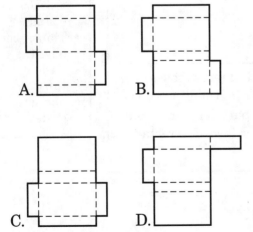

A.   B.

C.   D.

Solution

Method 1 Copy the four patterns onto paper or cardboard. Cut them out and fold on dotted lines to determine which would work and which would not work.

Method 2 Visualize the folds and the relationships of the different faces.

Answer Choice A would not make a closed box because opposite faces would not turn out to be congruent.

2. A plane passing through a solid gives a cross section of the solid. Determine the cross section of a cone if a plane passes through the cone parallel to the base.

Solution You can combine the strategy of drawing a diagram with the strategy of thinking about a physical model.

Think of a traffic cone being cut by a piece of cardboard parallel to the base. What do you see when the vertex has been cut off the cone?

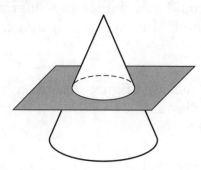

Answer The cross section of the cone is a circle.

3. Use cubes to make a model based on the views provided, and then draw a corner view.

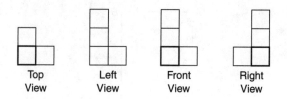

| Top View | Left View | Front View | Right View |

Solution If you use cubes to represent the views, then the top view tells us that the base has 3 blocks and that the figure has columns of different heights. The left view shows blocks flush with the surface. The left back must be three blocks high. From the front and right views it is clear the other columns are one block high. Hence, the figure should look like this:

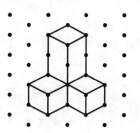

PRACTICE

1. Which of the following patterns will NOT fold to form a cube?

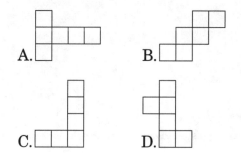

A. B.

C. D.

2. Which figure will be formed by folding the given pattern along the dotted lines?

A. hexagonal prism
B. hexagonal pyramid
C. cone
D. triangular prism

3. If rectangular region *ABCD* is rotated 360° about $\overline{BC}$, which of the following three-dimensional solids is formed?

A. cone
B. pyramid
C. prism
D. cylinder

4. Given the three views of the cube below, what shape is opposite the triangle?

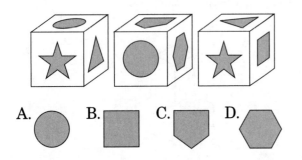

A. ⬤   B. ⬛   C. ⬠   D. ⬡

5. How many cubes are needed to build the given figure?

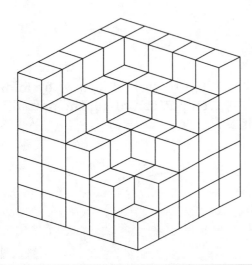

6. Explain why the diagram shown cannot be folded along the dotted lines to make a closed box.

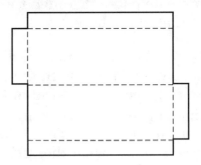

How could you change the diagram so that the folding will result in a closed box?

7. Draw the solid figure described by the given views.

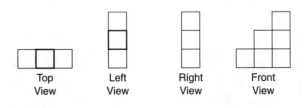

Top View   Left View   Right View   Front View

8. For the given solid, sketch the top, left, right, and front views.

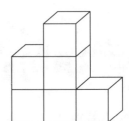

Assessment Macro A

1. Two lines are cut by a transversal as shown. As a result, which of the following statements is NOT necessarily true?

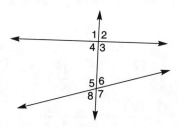

A. ∠1 is supplementary to ∠2.
B. ∠1 is supplementary to ∠4.
C. ∠5 ≅ ∠7
D. ∠6 ≅ ∠4

2. Arrange the three solid shapes in order of the number of vertices they contain, from LEAST to GREATEST.

 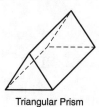 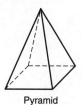

Cube   Triangular Prism   Pyramid

A. pyramid, cube, prism
B. cube, prism, pyramid
C. pyramid, prism, cube
D. prism, pyramid, cube

3. If *ABCDE* is a regular pentagon, what is the value of $x + y$?

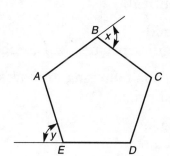

A. 72°
C. 120°
B. 108°
D. 144°

4. Which of the following is TRUE?

A. Every square is a rhombus.
B. Every rectangle is a square.
C. Every rhombus is a square.
D. Every parallelogram is a rectangle.

5. The diagram shows how many angles being bisected?

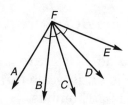

A. 3
C. 6
B. 4
D. 8

6. *C* is the midpoint of line segment $\overline{AB}$.
D is the midpoint of line segment $\overline{AC}$.
E is the midpoint of line segment $\overline{CB}$.
If $\overline{AD}$ = 6 inches, how long is $\overline{CB}$?

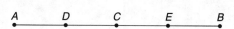

A. 6 inches
B. 12 inches
C. 18 inches
D. cannot be determined

7. Rita is going to use four straws to form quadrilaterals. If two of the straws are each 6 inches long and the other two are each 8 inches long, which of the following quadrilaterals will she NOT be able to form?

A. rectangle
C. rhombus
B. parallelogram
D. trapezoid

8. In the figure shown:

$\overline{AC}$ is parallel to $\overline{DF}$.
$\overline{AF}$ is parallel to $\overline{BE}$.

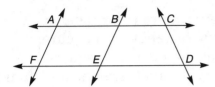

Based on the information given, which of the following is TRUE?

A. *ABEF* and *BCDE* are parallelograms.
B. *ACDF* and *BCDE* are trapezoids.
C. *BCDE* is an isosceles trapezoid.
D. *ABEF* is a rhombus.

9. If you spin this two-dimensional figure about the axis shown, you will generate a solid of revolution. Which solid would you generate?

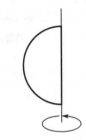

A.   B.

C.   D.

10. Which of the following patterns folds to give the cube pictured?

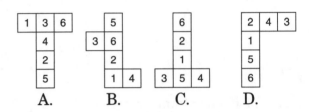

A.   B.   C.   D.

11. Which pattern can be folded to form a cylinder?

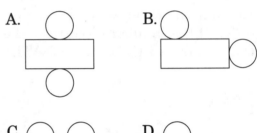

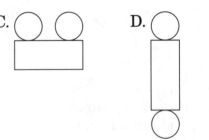

12. In which of the following solids are the top view, front view, and right-side view identical?

A. pyramid   B. cylinder
C. triangular prism   D. cube

13. Which of the following are faces of an octagonal prism?

  I. octagon
 II. triangle
III. rectangle

A. I only   B. I and II
C. II and III   D. I and III

14. What is the sum of the measure of the interior angles of a polygon with 32 sides?

15. The figure contains two squares and an equilateral triangle. What is the degree measure of $\angle x$?

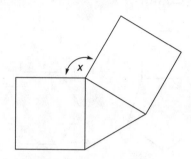

16. Lines l and m are parallel. If the measure of $\angle a = 115$, how many angles in the diagram measure 65°?

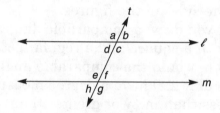

A. 0   B. 2   C. 3   D. 4

17. A triangle has vertices at (0, 0), (6, 0), and (3, 10). Which of the following best describes this triangle?

A. acute isosceles triangle
B. right isosceles triangle
C. obtuse isosceles triangle
D. acute scalene triangle

18. *ABCDEF* is a regular hexagon. Three diagonals are drawn as shown. What is the measure of $\angle FBD$?

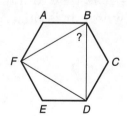

19. In pentagon *ABCDE*, diagonals $\overline{AC}$ and $\overline{AD}$ are drawn. The measures of $\angle B$ and $\angle E$ are each 100°. What is the sum of the measures of $\angle a$, $\angle b$, $\angle c$, $\angle d$, $\angle e$, $\angle f$, and $\angle g$?

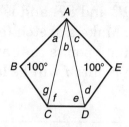

20. Which of the following is NOT true about a regular pentagon?

A. Each angle is an obtuse angle.
B. The sides are congruent.
C. The figure contains ten diagonals.
D. The sum of the interior angles is 540°.

21. Which of the following cubes could be made by folding the net shown?

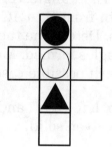

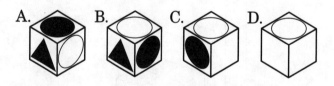

22. Find the measure of the angle made by the hands of a clock at 9:30.

23. $\overline{LN}$ has midpoint M and is parallel to $\overline{QR}$ and $\overline{ST}$. $\overline{LN}$ is located halfway between $\overline{QR}$ and $\overline{ST}$, and is 3 units from each. Make a sketch to figure out how many points that are 4 units away from point M will be located on $\overline{QR}$ or $\overline{ST}$.

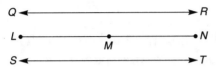

24. Explain why a pentagon cannot exist with the following interior angles: one interior angle measures 100° and each additional interior angle measures 15° more than the previous one.

25. In right scalene triangle ABC, B is the vertex of the right angle. A line segment is drawn from B to $\overline{AC}$ forming two triangles. Determine the possible types of triangles formed. Make a drawing of each possible case.

26. Draw the top, left, right, and front views for the given solid.

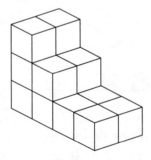

27. The figure on the left shows a square and a regular pentagon. The figure on the right shows a square and a regular hexagon.

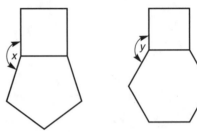

a. Find m∠x and m∠y. Show your process.
b. Explain why you can find these values without knowing the lengths of the sides of the figures.
c. If you draw a comparable diagram with a square and a regular octagon, would the comparable angle (call it ∠z) have a degree measure less than m∠y or greater than m∠y? Explain.

28. Examine the following statements about rectangles.

    A. The opposite sides are parallel.
    B. The diagonals are perpendicular.
    C. The opposite sides are congruent.
    D. The diagonals are congruent.

a. Which of the above statements is NOT true for all rectangles?
b. Explain your answer by showing a sketch of a rectangle where the statement is true and another sketch where it is false.

29. Emmanuel built the cube shown from smaller cubes. She then painted the exterior surface of the large cube red. When the paint dried, she separated the large cube into the 27 small cubes.

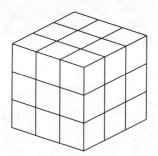

a. Copy and complete the table to show the number of small cubes containing each number of red faces.

Number of Red Faces	Number of Small Cubes
0	
1	
2	
3	
4	
5	
6	

b. If a large cube is built from 64 small congruent cubes and the exterior surface of the large cube is painted red, how many of the small cubes will not have red paint on any faces?

Apply the principles of congruence, similarity, symmetry, geometric transformations, and coordinate geometry.

2 B 1 Congruence

Congruent figures have the same size and the same shape. The symbol for congruence is ≅.

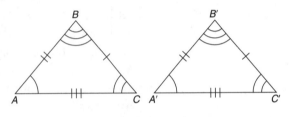

$$\triangle ABC \cong \triangle A'B'C'$$

$\angle A \cong \angle A'$	$\overline{BC} \cong \overline{B'C'}$
$\angle B \cong \angle B'$	$\overline{AC} \cong \overline{A'C'}$
$\angle C \cong \angle C'$	$\overline{AB} \cong \overline{A'B'}$

Characteristics of Congruent Polygons

1. Corresponding angles are congruent.
2. Corresponding sides are congruent.

Model Problem

Give a set of coordinates for point Q so that parallelogram $ABCD$ will be congruent to parallelogram $MNPQ$.

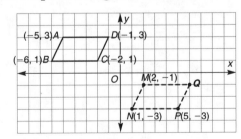

Solution   If parallelogram $ABCD \cong$ parallelogram $MNPQ$, $\overline{AD} \cong \overline{MQ}$. Since $\overline{AD} = 4$ units, $\overline{MQ}$ must be 4 units.

Answer   Point Q must be located at $(6, -1)$.

1. Quadrilateral *ABCD* is congruent to quadrilateral *PQRS*. Which statement does NOT follow?

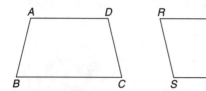

   A. $\overline{AB} \cong \overline{PQ}$ B. $\overline{AD} \cong \overline{RS}$
   C. $\overline{BC} \cong \overline{QR}$ D. $\overline{CD} \cong \overline{RS}$

2. Which of the following everyday applications is an illustration of congruence?

   A. obtaining an enlargement of a picture
   B. duplicating a key
   C. making a scale drawing
   D. building a scale model of a sports car

3. $\triangle DAB \cong \triangle CBA$. Which angle is congruent to $\angle BDA$?

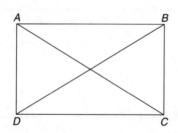

   A. $\angle ACB$ B. $\angle BAC$
   C. $\angle ABC$ D. $\angle CAB$

4. Which of the following statements are TRUE?

   I. If two circles have the same area, the circles are congruent.
   II. If two squares have the same area, the squares are congruent.
   III. If two rectangles have the same area, the rectangles are congruent.

   A. II only   B. III only
   C. I and II only   D. II and III only

5. $\triangle ABC \cong \triangle EFG$, $m\angle A = 40$, $m\angle F = 105$. What is the measure of $\angle C$?

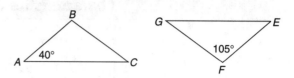

6. What is the *x*-coordinate of point *Z* such that trapezoid *ABCD* $\cong$ trapezoid *WXYZ*?

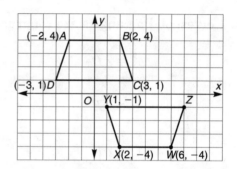

7. Take a regular octagon. Draw all the diagonals from one vertex. How many pairs of congruent triangles are formed?

8. Roman thinks that if the angles of one triangle are congruent to the angles of another triangle, then the two triangles must be congruent. Do you agree with Roman? Explain. Give an example to support your answer.

9. Draw two triangles that meet each of the given conditions. Be sure to label the dimensions on the triangles.

   a. The two triangles have equal areas and are congruent.
   b. The two triangles have equal areas and are not congruent.

2 B 2 Similarity

Similar figures have the same shape, but not necessarily the same size. The symbol for similarity is ~.

Characteristics of Similar Polygons

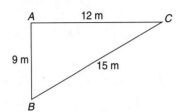 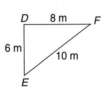

$$\triangle ABC \sim \triangle DEF$$

1. Corresponding angles are congruent.

$$\angle A \cong \angle D$$

$$\angle B \cong \angle E$$

$$\angle C \cong \angle F$$

2. Corresponding sides are proportional.

$$\frac{BC}{EF} = \frac{15 \text{ m}}{10 \text{ m}} = \frac{3}{2}$$

$$\frac{AC}{DF} = \frac{12 \text{ m}}{8 \text{ m}} = \frac{3}{2}$$

$$\frac{AB}{DE} = \frac{9 \text{ m}}{6 \text{ m}} = \frac{3}{2}$$

The reduced ratio, $\frac{3}{2}$, is called the **ratio of similitude**.

Model Problem

1. Quadrilateral $ABCD \sim$ quadrilateral $PQRS$. Find $x, y,$ and z.

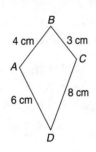 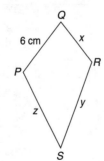

Solution   In similar quadrilaterals, corresponding sides are in proportion.

$$\frac{AB}{PQ} = \frac{BC}{QR} \qquad \frac{AB}{PQ} = \frac{CD}{RS} \qquad \frac{AB}{PQ} = \frac{AD}{PS}$$

$$\frac{4}{6} = \frac{3}{x} \qquad\qquad \frac{4}{6} = \frac{8}{y} \qquad\qquad \frac{4}{6} = \frac{6}{z}$$

$$4x = 18 \qquad\qquad 4y = 48 \qquad\qquad 4z = 36$$

Answer $x = 4.5$ cm $y = 12$ cm $z = 9$ cm

Note: Since $\overline{PQ}$ and $\overline{AB}$ are corresponding sides and the length of $\overline{PQ}$ is one and one-half times the length of $\overline{AB}$, the missing sides in the larger quadrilateral must be one and one-half times the corresponding sides in the smaller quadrilateral.

2. A person 6-feet tall is standing near a tree. If the person's shadow is 4-feet long and the tree's shadow is 10-feet long, what is the height of the tree?

Solution   During the day, two objects that are near each other have shadows whose measures are proportional to the heights of the objects, resulting in similar triangles.

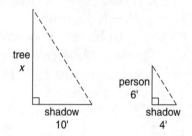

Write a proportion and solve it:

$$\frac{\text{tree}}{\text{tree's shadow}} = \frac{\text{person}}{\text{person's shadow}}$$

$$\frac{x}{10} = \frac{6}{4}$$

$$4x = 60$$

$$x = 15$$

Answer   The tree is 15 feet high.

PRACTICE

1. Which of the following statements about similar figures is TRUE?

   A. All squares are similar.
   B. All right triangles are similar.
   C. All rhombuses are similar.
   D. All hexagons are similar.

2. $\triangle ABC \sim \triangle DEF$. What is the measure of $\overline{DF}$?

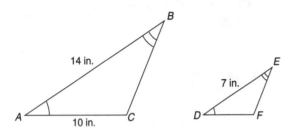

   A. 5 in.   B. 11 in.
   C. 17 in.   D. 20 in.

3. Rectangle $ABCD \sim$ rectangle $PQRS$. If $AD = 3$ mi and $PS = 5$ mi, which of the quantities that follow would NOT be in the ratio 3:5?

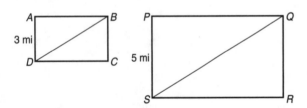

   A. AB and PQ
   B. BD and QS
   C. $m\angle A$ and $m\angle P$
   D. the perimeter of $ABCD$ and the perimeter of $PQRS$

4. A tree casts a 20-meter shadow at the same time that a 6-meter pole casts an 8-meter shadow. Find the height of the tree.

5. Each wing of a model of a glider is a right triangle with sides 3 inches, 4 inches, and 5 inches. In the actual glider, the hypotenuse is 35 feet. Find the perimeter of the wing on the actual glider.

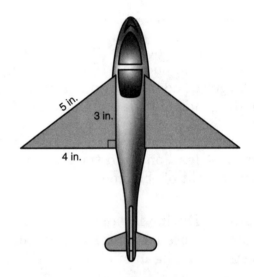

6. Triangle ABC has vertices $A(-4, 1)$, $B(-1, 1)$, and $C(-4, 3)$. Triangle PQR is to be drawn similar to the first triangle. If two vertices of the second triangle are $P(2, 1)$ and $Q(8, 1)$, find the positive y-coordinate for the vertex R if the x-coordinate is 8.

7. If the two triangles shown are similar, explain why $m\angle x$ must be equal to $m\angle y$.

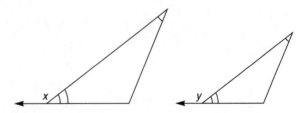

8. $\triangle ABC$ has vertices $A(0, 0)$, $B(6, 0)$, and $C(3, 3)$. If the coordinates of each vertex are multiplied by 2, will the new

figure be similar to the original figure? Give the coordinates of the new vertices. Plot the coordinates and draw both triangles on the coordinate plane.

9. Jennifer knows that two rectangles are similar and that the ratio of corresponding sides is 2:5. She also knows that the ratio of the perimeters of the two rectangles is 2:5. Based on this knowledge, Jennifer thinks that the ratio of the areas of the two rectangles is 2:5. Using graph paper, draw approximate rectangles to prove or disprove Jennifer's thinking.

2 B 3 Transformations

Reflection and Line Symmetry

When you look in a mirror, you see your reflected image. A **reflection** is a flipping of a geometric figure about a line to obtain its mirror image.

Note that under a reflection, the **orientation** of the figure, the way it is facing, is changed. For example, point A is to the left of B in the original figure, or **preimage**, but, in the image, point A' is to the right of B'.

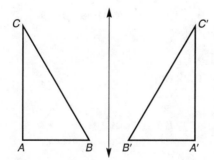

If a line can be drawn through a figure so that the part of the figure on one side of the line is the mirror image of the part on the other side of the line, the figure has **line symmetry**.

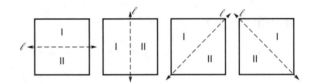
Model Problem

1. Draw the reflection of $\triangle ABC$ over the x-axis. State the coordinates of the vertices of the image.

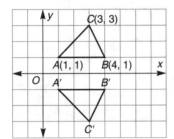

Solution   Point A has coordinates $(1, 1)$. Since point A is 1 unit above the reflecting line, point A' will be 1 unit below the line. Since point $B(4, 1)$ is also 1 unit above the reflecting line, point B' will be 1 unit below the line. Since point $C(3, 3)$ is 3 units above, point C' will be 3 units below.

Answer   point $A(1, 1) \rightarrow$ point $A'(1, -1)$
point $B(4, 1) \rightarrow$ point $B'(4, -1)$
point $C(3, 3) \rightarrow$ point $C'(3, -3)$

2. How many lines of symmetry does a square have?

Solution

In each of the figures shown, line l is a line of symmetry. If the square is folded on line l, then region I will coincide exactly with region II.

Answer   There are four lines of symmetry.

Rotation

A **rotation** is a transformation that turns a figure about a point. To rotate a figure, you must have:

- a center of rotation about which to rotate the figure.
- a direction of rotation (clockwise or counterclockwise).
- the number of degrees of rotation.

In the figure shown, C is the center of rotation, the direction is clockwise, and the rotation is 90°.

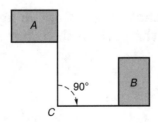

Rotate $\triangle ABC$ 90° clockwise using (0, 0) as the point of rotation. Identify the coordinates of the vertices of the image.

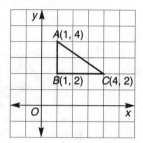

Solution   Trace $\triangle ABC$ on graph paper. Locate (0, 0) and draw a ray away from the origin and along the negative branch of the x-axis. Hold the paper with pencil point at (0, 0) and rotate the paper 90° clockwise, moving the ray from the x-axis to the y-axis.

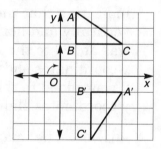

$$A(1, 4) \rightarrow A'(4, -1)$$
$$B(1, 2) \rightarrow B'(2, -1)$$
$$C(4, 2) \rightarrow C'(2, -4)$$

Translation

A **translation** is a sliding of a geometric figure, without turning, from one position to another.

1. Translate $\triangle ABC$ right 3 units. State the coordinates of the vertices of the image of $\triangle ABC$.

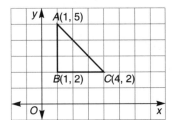

Solution The image of A under the translation 3 units right is A'. Add 3 to the x-coordinate of point A.

$$A(1, 5) \rightarrow A'(1 + 3, 5) \rightarrow A'(4, 5)$$

Repeat for points B and C

$$B(1, 2) \rightarrow B'(4, 2)$$
$$C(4, 2) \rightarrow C'(7, 2)$$

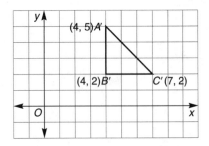

2. Explain the translation used to produce the given image of trapezoid $ABCD$.

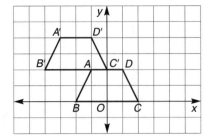

Solution $\quad A(-1, 2) \rightarrow A'(-3, 4)$
$\qquad\qquad\quad B(-2, 0) \rightarrow B'(-4, 2)$
$\qquad\qquad\quad C(2, 0) \rightarrow C'(0, 2)$
$\qquad\qquad\quad D(1, 2) \rightarrow D'(-1, 4)$

Both the x- and y-coordinates changed, indicating that the figure slid both vertically and horizontally. Since -2 was added to the x-coordinates, the x-coordinates slid 2 units left. Since 2 was added to the y-coordinates, the y-coordinates slid 2 units up.

Answer The image of trapezoid $ABCD$ was formed by translating the figure left 2 units and up 2 units.

Dilation

A **dilation** is a transformation that reduces or enlarges a figure. In a reflection, rotation, or translation, the preimage and the image are congruent. In a dilation, the side lengths of the figure change, while the angle measures remain the same. Therefore, in a dilation every preimage is *similar* to its image.

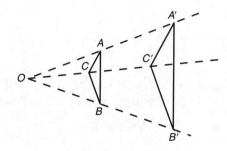

Every dilation has a point known as the **center of dilation** and a scale factor, k. The center of dilation is usually the origin, but can be any point. The scale factor, k, is the ratio between corresponding sides of the preimage and the image. If $k > 1$, the dilation is an **enlargement**. If $0 < k < 1$, the dilation is a **reduction**. If k is expressed as a fraction, it represents the ratio of similitude between the image and the preimage.

 # Model Problem

1. Using the diagram, identify the dilation and find its scale factor.

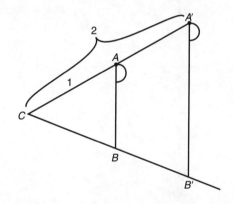

Solution $\triangle A'B'C$ is a dilation of $\triangle ABC$. To find the scale factor, write the ratio between corresponding sides CA' and CA.

$$k = \frac{CA'}{CA} = \frac{2}{1} = 2$$

Answer The dilation is an enlargement because the scale factor, 2, is greater than 1.

2. Draw a dilation of rectangle $PQRS$ with vertices $P(2, 2)$, $Q(4, 2)$, $R(4, 3)$, and $S(2, 3)$. Use the origin as the

center and a scale factor of 3. How does the perimeter of the preimage compare with the perimeter of the image?

Solution With center at $(0, 0)$ you find the image of each vertex by multiplying its coordinates by the given scale factor.

$$P(2, 2) \rightarrow P'(6, 6)$$
$$Q(4, 2) \rightarrow Q'(12, 6)$$
$$R(4, 3) \rightarrow R'(12, 9)$$
$$S(2, 3) \rightarrow S'(6, 9)$$

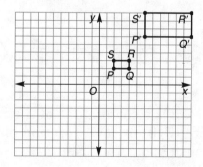

Answer From the graph we find that the perimeter of $PQRS = 6$ and the perimeter of $P'Q'R'S' = 18$. The perimeter was enlarged by a scale factor of 3.

1. A figure undergoes a transformation. The preimage and the image are not congruent. What type of transformation took place?

   A. dilation
   B. reflection
   C. rotation
   D. translation

2. What would the coordinate of point A' be if $\triangle ABC$, with vertices at $A(1, 1)$, $B(4, 1)$, and $C(2, 3)$, was translated left 3 units?

   A. $(4, 1)$
   B. $(-2, 1)$
   C. $(-2, 2)$
   D. $(1, 2)$

3. If the line segment joining $A(-2, 3)$ and $B(1, 6)$ is rotated 90° clockwise about point A, the coordinates of the image of point B are:

   A. $(1, 0)$
   B. $(1, 6)$
   C. $(-2, 0)$
   D. $(-2, 6)$

4. Which of the following figures has NO lines of symmetry?

   A.   B.   C.   D.

5. What are the coordinates of the image of point $(5, 0)$ under a rotation of 90° clockwise about the origin?

   A. $(-5, 0)$
   B. $(0, 5)$
   C. $(0, -5)$
   D. $(5, -5)$

6. Which of the following could be the image of $\triangle ABC$ reflected over the x-axis?

   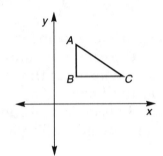

   A.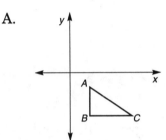

   B.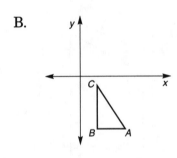

   C.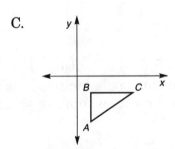

   D.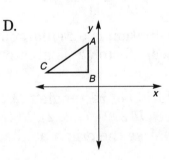

7. Which figure below could be a translation of $\triangle ABC$?

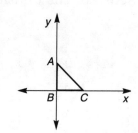

A.

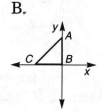

B.

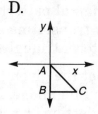

C.

D.

8. If $\triangle ABC$ with vertices $A(-2, 0)$, $B(-2, -2)$, and $C(-4, -2)$ is reflected over the x-axis and then the image is reflected over the y-axis, which coordinates would represent the final image of point A?

A. $(-2, 0)$
B. $(2, 0)$
C. $(0, 2)$
D. $(0, -2)$

9. If the x-axis is the line of reflection, what would be the coordinates of the reflected image of $P(6, 1)$?

A. $(6, -1)$
B. $(-6, 1)$
C. $(-6, -1)$
D. $(6, 1)$

10. Which of the following could be the reflected image of **B** over a vertical line?

A. **ꓭ**   B. **ꓭ**   C. **B**   D. **ꓭ**

11. Which figure could be a translation of **R**?

A. **ꓤ**   B. **ꓩ**   C. **R**   D. **Я**

12. Which figure shows the preimage E in Quadrant II and the image of E after a rotation?

A.   B.   C.   D.

13. When the word MATH is written in column form, it has a vertical line of symmetry. Find two other words that have a vertical line of symmetry when written in column form.

14. If $\triangle RST$ is translated up 6 units, what will be the y-coordinate of point S'?

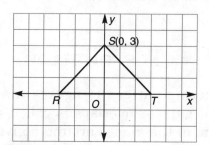

15. Parallelogram $ABCD$ was translated down 2 units and right 5 units, resulting in the given image. What was the x-coordinate of point B?

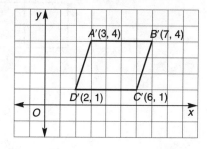

16. The vertices of $\triangle ABC$ are $A(-1, -2)$, $B(0, 3)$, and $C(1, 1)$. The image of A under a translation is $A'(-3, 5)$. What would the images of B and C be under the same translation?

17. On a coordinate plane, graph the given vertices: $A(2, 3)$, $B(4, 8)$, and $C(-3, 5)$. Using the origin as the center of the dilation and a scale factor of 2, draw the dilation image.

18. $\triangle ABC$ transformed into $\triangle A'B'C'$. Find the scale factor and determine if the dilation is a reduction or an enlargement.

$AB = 6$ cm $\quad BC = 8$ cm $\quad AC = 10$ cm
$A'B' = 24$ cm $B'C' = 32$ cm $A'C' = 40$ cm

19. $AB = 5$ mm. The measure of its dilated image is 7.5 mm. What is the value of the scale factor?

20. If the figure on the right is a dilation image of the figure on the left, what is the scale factor for the dilation?

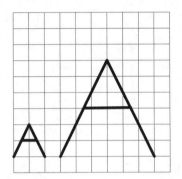

21. Wrapping paper often consists of a design that is translated over and over to create a pattern. Create an original pattern for wrapping paper by translating a hexagon and an equilateral triangle.

22. Plot the given points to form a figure and its image. Is this an example of a reflection? Explain why or why not.

figure: $A(-2, 9)$, $B(-2, 2)$, $C(-5, 2)$
image: $A'(9, 2)$, $B'(2, 2)$, $C'(2, 5)$

---

2 B 4 Tessellation

A repeating pattern of figures that completely covers a plane region without gaps or overlaps is called a **tessellation**. In any tessellation, the sum of the angles around a given point is 360°. For example, a tessellation can be formed using a square.

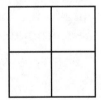

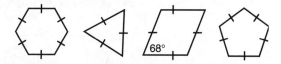

Model Problem

Which of the following figures will produce tessellations? Use drawings to support your answer.

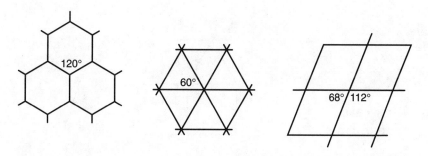

Solution The hexagon, the triangle, and the rhombus will produce tessellations.

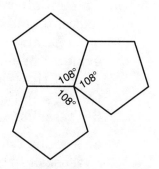

The regular pentagon will not produce a tessellation. Each interior angle of the pentagon is 108°, which does not go into 360° evenly. Therefore, pentagons arranged around a common vertex will not fill the plane, but will leave a gap as shown.

PRACTICE

1. If n represents the number of sides of a regular polygon, which value of n indicates a polygon that will NOT form a tessellation?

   A. 3   B. 4
   C. 5   D. 6

2. A dodecagon (12-sided polygon) cannot tessellate by itself. Which of the following figures can be used in combination with the dodecagon to create a tessellation?

   A. equilateral triangle
   B. square
   C. regular hexagon
   D. regular octagon

3. The given tessellation illustrates which of the following transformations?

   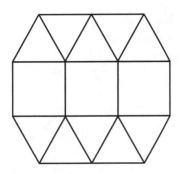

   A. reflection
   B. rotation
   C. translation
   D. dilation

4. The given tessellation uses equilateral triangles and regular hexagons. Find a different combination of regular polygons that can be used to create a tessellation where each polygon has an even number of sides. Draw a sketch of the new tessellation.

   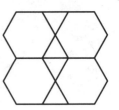

5. Which of the following patterns will NOT tessellate a plane?

   A.   B.

   C.   D.

6. An equilateral triangle can be used to create a tessellation. To find out if other triangles create tessellations, explore the following cases. In each case, if the triangle tessellates, show the tessellation. If the triangle does not tessellate, explain why.

   a. scalene triangle
   b. isosceles triangle
   c. right triangle
   d. Based on your findings, what conclusion can you come to about triangles and tessellations?

Assessment Macro B

1. Which of the following figures has exactly one line of symmetry?

A.   B.   C.   D.

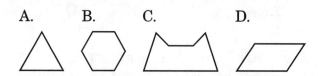

2. A triangle has sides of lengths 6 inches, 8 inches, and 9 inches. Which of the following would NOT be the lengths (in inches) of a similar triangle?

   A. $2, 2\frac{1}{3}, 3$
   B. 3, 4, 4.5
   C. 12, 16, 18
   D. 18, 24, 27

3. Three vertices of a rectangle are (2, 0), (6, 0), and (6, 3). What are the coordinates of the fourth vertex?

   A. (6, −3)
   B. (0, 3)
   C. (2, 3)
   D. (2, −3)

4. Which of the following statements is NOT true?

   A. Squares with the same perimeter are congruent.
   B. Any two equilateral triangles must be congruent.
   C. Any two isosceles right triangles must be similar.
   D. Squares that are not congruent squares have unequal areas.

5. The coordinates of point P are (2, −3). If you reflect P about the x-axis, what are the coordinates of the image P'?

   A. (2, −3)
   B. (−2, −3)
   C. (2, 3)
   D. (−2, 3)

6. Triangle ABC is translated so that the image is completely contained in the third quadrant. Which of the following could NOT have been the translation?

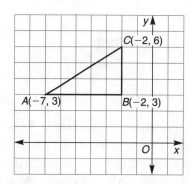

   A. down 8 units
   B. to the right 1 unit and down 10 units
   C. to the right 10 units and down 10 units
   D. to the left 6 units and down 10 units

7. Right triangle ABC, with the right angle at vertex C, is reflected about the x-axis. Which of the following properties of triangle ABC is changed as a result of the reflection?

   A. the measure of angle A
   B. the orientation of the triangle
   C. the length of the hypotenuse
   D. the area of the triangle

8. Under a translation, which of the following properties of a figure are preserved?

   I. the length of the sides
   II. the measure of the angles
   III. the orientation of the figure

   A. I only
   B. II only
   C. I and II
   D. I, II, and III

9. Which set of transformations on the white figure below will NOT make the black figure become the image of the white figure?

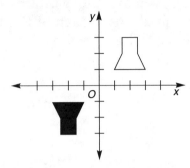

A. reflection over the *y*-axis followed by reflection over the *x*-axis
B. translation 4 units left followed by reflection over the *x*-axis
C. translation 4 units left followed by translation 2 units down
D. rotation 180° clockwise about the origin

10. If a triangle in the second quadrant is reflected over the *x*-axis and its image is next reflected over the *y*-axis, in which quadrant will the final image appear?

A. I   B. II
C. III   D. IV

11. $\triangle XYZ$ is similar to $\triangle RST$. What is the length of $\overline{ST}$?

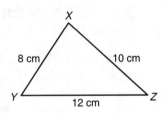

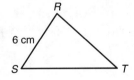

A. 5 cm   B. 7.5 cm
C. 9 cm   D. 10 cm

12. $\triangle PQR$, with vertices $P(3, 2)$, $Q(7, 2)$, and $R(5, -1)$, is reflected over the *y*-axis and then the image is translated 3 units to the right. Which coordinates would represent the final image of point *R?*

A. $(-8, 1)$ B. $(-5, 1)$
C. $(-2, -1)$ D. $(8, 1)$

13. $\triangle ABC$ has a line segment on the *x*-axis as its base and a point on the *y*-axis as its third vertex. The reflection of $\triangle ABC$ over the *y*-axis causes the image to coincide with the original triangle.

For which of the following triangles would this relationship be true?

   I. equilateral triangle
  II. isosceles triangle
 III. scalene triangle

A. I only   B. III only
C. I and II   D. I, II, and III

14. The point $(5, 6)$ is reflected over the *y*-axis and the image point is then translated 3 units to the left. What are the coordinates of the final image?

A. $(-5, 6)$ B. $(-8, 6)$
C. $(-2, 6)$ D. $(2, 6)$

15. Which of the following does NOT show a pair of similar triangles?

A.   B.

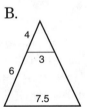

C.   D.

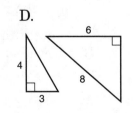

16. Jack wants to use an isosceles triangle with an 18° vertex angle to make a tessellation. How many triangles will have to meet at a vertex to complete the tessellation?

17. A tree casts a shadow 20 meters long at the same time that a man 2 meters tall casts a shadow of 5 meters. What is the height of the tree?

18. A ramp that is 50 yards long slopes up from the ground to a height of 15 yards. A vertical support beam is to be built 20 yards from the top of the ramp. How long (in yards) will the support beam be?

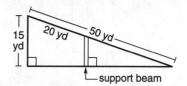

Open-Ended Questions

19. A triangular plot of land is represented by the diagram.

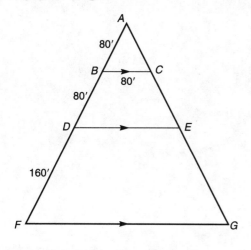

$$\overline{AB} = 80 \text{ feet}$$
$$\overline{BD} = 80 \text{ feet}$$
$$\overline{BC} = 80 \text{ feet}$$
$$\overline{DF} = 160 \text{ feet}$$

Fences are going to be built along the parallel segments $\overline{BC}$, $\overline{DE}$, and $\overline{FG}$. What is the total length of fencing (in feet) needed for the three sections?

20. Given the coordinates $A(-4, 0)$, $B(4, 0)$, $C(1, 8)$, and $D(-1, 8)$:

a. Plot the points and form a figure.
b. Give the best name for the figure formed.
c. Find the area of the figure. Show the steps involved in your process.

21. Two isosceles triangles have unequal areas. Explain why this is NOT enough information to determine if the two triangles are similar. Draw a sketch to illustrate your explanation.

22. An equilateral triangle has three lines

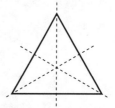

of symmetry as shown.

a. Sketch regular polygons with 4, 5, 6, and 8 sides.
b. Complete the following table to

Number of sides	3	4	5	6	8
Number of lines of symmetry	3	?	?	?	?

indicate the number of lines of
symmetry for each regular polygon.

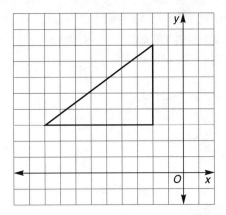

23. Triangle *ABC* is located in the second
quadrant as shown.
a. Show the image when triangle *ABC* is
reflected over the *y*-axis, and then
reflected again over the *x*-axis.
b. What single transformation of triangle
ABC would give the same result as the
two reflections in part a above? Explain

Apply the principles of measurement and geometry to solve problems involving direct and indirect measurement.

2 C 1 Perimeter and Circumference

The **perimeter** of a figure is the total distance around the outside of the figure. Perimeter is measured in linear units (such as centimeters, inches, feet, and meters).

The distance around the outside of a circle is called the **circumference** instead of the perimeter.

Formulas can be used to find the perimeters of common geometric figures.

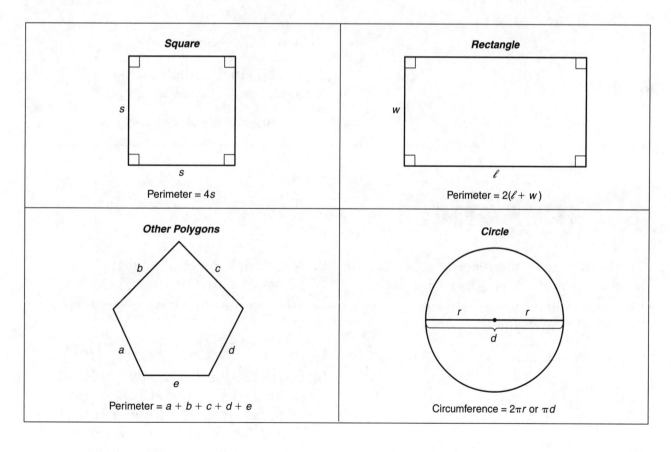

Square	**Rectangle**
Perimeter = 4s	Perimeter = 2(ℓ + w)
Other Polygons	**Circle**
Perimeter = a + b + c + d + e	Circumference = 2πr or πd

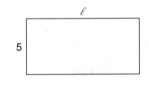

Model Problem

1. A rectangle with a width of 5 cm has the same perimeter as an equilateral triangle with a side of 12 cm. Find the length of the rectangle.

Solution Draw a diagram to help.

Perimeter is the distance around the outside.

$$P_{triangle} = 12 + 12 + 12 = 36$$
$$P_{rectangle} = 2(l + w)$$
$$P_{triangle} = P_{rectangle}$$
$$36 = 2(l + 5)$$
$$36 = 2l + 10$$
$$26 = 2l$$
$$13 = l$$

Answer   13 cm

2. A circle is inscribed in a square with a perimeter of 24 cm. What is the circumference of the circle?

Solution Draw a picture to help with the solution.

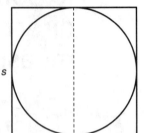

$$P_{square} = 4s$$
$$24 = 4s$$
$$6 = s$$

A side of the square is 6 cm. The length of a diameter of the circle is equal to the length of a side of the square.

$$C = \pi d$$
$$C = \pi(6)$$

Answer $C = 6\pi$ cm

Note: If a single-number answer is needed, use $\pi = 3.14$ to simplify.

$$6\pi = 6(3.14) = 18.84 \text{ cm}$$

PRACTICE

1. The perimeter of the rectangle *ABCD* is 40 units and its length is 15 units. Find the number of units in the perimeter of the square.

A. 20   B. 25   C. 40   D. 60

2. A rectangle has three vertices at points $(-2, 3)$, $(10, 3)$, and $(10, 6)$. What is the perimeter of the rectangle?

A. 22   B. 26   C. 30   D. 36

3. David leashes his puppy Chelsea to a stake in the ground. Chelsea can run in circles around the stake in a path that is 30π feet around. David thinks the puppy needs more exercise and wants her to be able to run 50π feet around. How much longer should David make Chelsea's leash?

A. 10 feet   B. 10π feet
C. 20 feet   D. 20π feet

4. Six congruent squares are shown. Squares added to the figure must share an existing side. What is the minimum number of squares that need to be added to the diagram in order to increase the perimeter to 18 units?

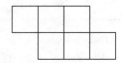

5. A regular pentagon and a regular decagon each have sides with a length of 8 cm. What is the difference between the two perimeters?

6. The diagram contains 5 rectangles each with dimensions of 4 cm by 1 cm. What is the perimeter of the entire figure?

A. 20 cm   B. 40 cm
C. 48 cm   D. 50 cm

7. A rectangle has perimeter of 60 inches and length of 20 inches. A second rectangle has dimensions of 20% more than the dimensions of the first rectangle. What is the perimeter of the second rectangle?

A. 72 inches   B. 80 inches
C. 144 inches   D. 240 inches

8. Elena is making a wall hanging to decorate her bakery. She is using a circular piece of plywood, as seen in the diagram. Elena wants to glue lace completely around the shaded part of the circle. How much lace will she need if the diameter of the plywood circle is 18 inches and the measure of the central angle is 120°? Round your answer to the nearest inch.

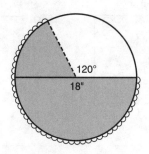

9. For the given figure, what happens to the total perimeter of the figure if you remove the squares as follows? Explain your answers.

1	2	3
4	5	6

a. Remove square 3 only.
b. Remove square 2 only.
c. Which other square could you remove and have the same effect as in part a?
d. Which other square could you remove and have the same effect as in part b?
e. Draw a diagram to show how you could add three squares to the original figure and thereby change the perimeter from the original 10 units to 14 units.

10. Use a centimeter ruler to determine the length of the radius of the given circle. Using this measurement, find the length of the 120° arc of the circle (to the nearest tenth of a centimeter). Explain how you go about finding the length of this arc.

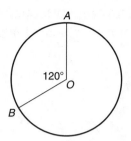

---

2 C 2 Area

Area is the number of square units needed to cover a surface.

Formulas can be used to find the areas of common geometric figures.

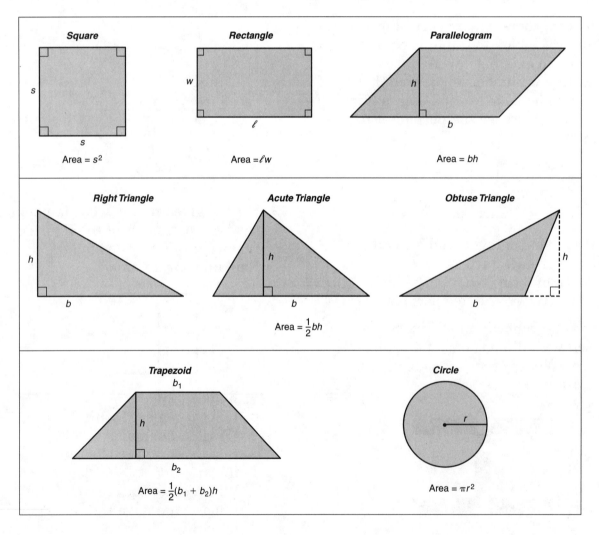

Model Problem

1. Arrange the three figures in INCREASING order of area.

    I. a rectangle with dimensions of 6 in. by 8 in.
   II. a square with a side of 7 in.
  III. a right triangle with sides of 5 in., 12 in., and 13 in.

    A. rectangle, square, triangle
    B. rectangle, triangle, square
    C. triangle, rectangle, square
    D. square, rectangle, triangle

Solution Find the areas by using the appropriate formulas.

$$\text{Area of rectangle} = lw$$
$$A = (6)(8)$$
$$A = 48 \text{ sq in.}$$

$$\text{Area of square} = s^2$$
$$A = 7^2$$
$$A = 49 \text{ sq in.}$$

$$\text{Area of triangle} = \frac{1}{2}\,bh$$
$$A = \frac{1}{2}\,(5)(12)$$
$$A = 30 \text{ sq in.}$$

The areas in increasing order are: triangle, rectangle, and square.

Answer   C

2. Use the floor plan pictured below to find the minimum number of square yards of carpeting needed to carpet the entire floor from wall to wall. Show your procedure.

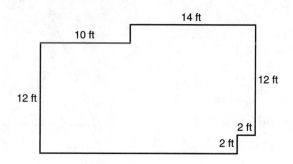

Solution One way to find the area of the floor is to divide the floor into rectangles.

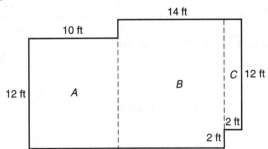

Rectangle A
$A = lw$
$A = (12)(10)$
$A = 120$ sq ft

Rectangle B
$A = lw$
$A = (12)(14)$
$A = 168$ sq ft

Rectangle C
$A = lw$
$A = (2)(12)$
$A = 24$ sq ft

$$\text{Total Area} = A + B + C$$
$$= 120 + 168 + 24$$
$$= 312 \text{ sq ft}$$

Since there are 9 square feet in a square yard, divide by 9 to find the square yards.

$$312 \text{ sq ft} = 34.66 \text{ sq yd}$$

Answer The minimum quantity of carpeting needed is 35 square yards.

PRACTICE

1. The area of a rectangle is 200 sq cm and one dimension is 25 cm. Find the length of a diagonal of the rectangle (to the nearest whole number).

   A. 8 cm   B. 26 cm
   C. 33 cm   D. 90 cm

2. If a radius of a circle is tripled, then the area is:

   A. increased by 3   B. multiplied by 9
   C. tripled   D. cubed

3. If you draw a square with an area equal to the area of triangle ABC, how long will each side of the square be?

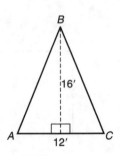

   A. a little less than 7 ft
   B. a little less than 10 ft
   C. a little more than 10 ft
   D. a little less than 14 ft

4. What is the area of the shaded portion of the triangle?

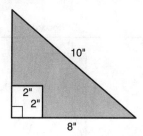

A. 8 sq in.   B. 20 sq in.
C. 36 sq in.   D. 44 sq in.

5. Which of the triangles has an area unequal to that of the other three triangles?

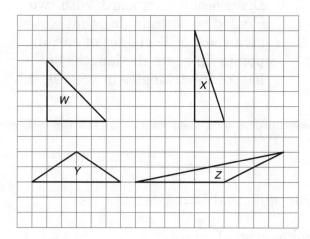

A. *W*   B. *X*
C. *Y*   D. *Z*

6. A circle has a diameter with endpoints at $(5, 0)$ and $(-5, 0)$. How many square units are in the area of the circle?

A. 5π B. 25
C. 25π D. 100π

7. In the figure, each side of the square has been divided into four equal segments. What is the ratio of the shaded area to the total area?

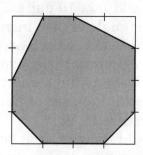

A. $\dfrac{3}{16}$ B. $\dfrac{13}{16}$ C. $\dfrac{3}{8}$ D. $\dfrac{5}{8}$

8. The sum of lengths of the two bases of a trapezoid is 20 inches. If the height of the trapezoid is 8 inches, what is the area of the trapezoid?

9. In a town, the property tax is $0.06 per square foot of land. What would the tax be on the plot of land pictured?

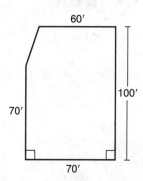

10. If there are 100 square units in the grid, what is the area, in square units, of the shaded figure?

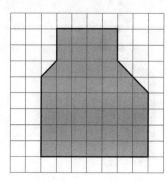

11. A rectangle has a length of 20 units and a width of 10 units. The length of the rectangle is then increased by 20% and the width is decreased by 10%.
    a. Explain how to determine what happens to the area in terms of percent increase or percent decrease.
    b. Would the new area be the same if the original length were increased 10% and the original width were decreased 20%? Explain your answer.

12. Players can earn points for darts landing in the shaded region (two right triangles) of the rectangular dartboard shown.

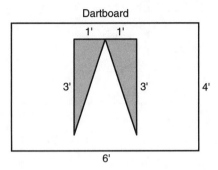

Dartboard

a. What portion of the dartboard is shaded? Express your answer as both a fraction and a percent.
b. Draw a new dartboard, with two shaded right triangles, that would give you a 25% chance of earning points. Indicate the lengths of the legs of the triangles in your sketch.

2 C 3 Volume

The **volume** of a solid figure is the number of cubic units that fit inside the solid.

Formulas can be used to find the volumes of common solid figures.

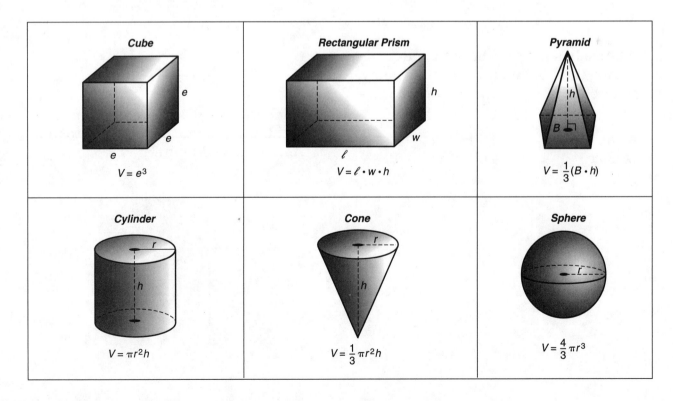

Cube	Rectangular Prism	Pyramid
$V = e^3$	$V = \ell \cdot w \cdot h$	$V = \frac{1}{3}(B \cdot h)$
Cylinder	Cone	Sphere
$V = \pi r^2 h$	$V = \frac{1}{3}\pi r^2 h$	$V = \frac{4}{3}\pi r^3$

![Rubik's cube icon] # Model Problem

1. Which of the following rectangular prisms does NOT have a volume of 48 cubic units?

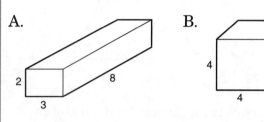

A.

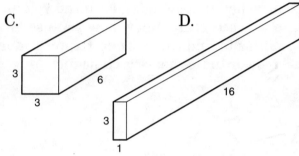

B.

C.

D.

2. The Air Puff Popcorn Company is considering the three designs shown as new containers for their popcorn.

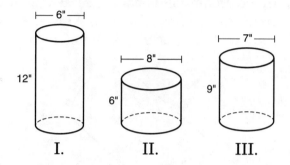

I.   II.   III.

Which container holds the most popcorn?

Solution   The container that holds the most popcorn is the one with the greatest volume.

$$V = \pi r^2 h$$

I. $V = 3.14(3)^2 \times (12) = 339.12$ in.^{3}
II. $V = 3.14(4)^2 \times (6) = 301.44$ in.^{3}
III. $V = 3.14(3.5)^2 \times (9) = 346.185$ in.^{3}

Answer   Container III holds the most popcorn.

Solution

Use V = area of base $\times$ height.

For a rectangular prism, $V = lwh$.
Prism A $V = 2 \times 3 \times 8 = 48$ cu units
Prism B $V = 4 \times 4 \times 3 = 48$ cu units
Prism C $V = 3 \times 3 \times 6 = 54$ cu units
Prism D $V = 3 \times 1 \times 16 = 48$ cu units

Answer   C

![Grid and pencil icon] # PRACTICE

1. If the edge of a cube is doubled, the volume is multiplied by:

A. 2   B. 3
C. 6   D. 8

2. Which of the following has a volume different from the other three volumes?

A. a cylinder with a radius of 4 cm and a height of 9 cm
B. a cylinder with a radius of 5 cm and a height of 6 cm
C. a cylinder with a radius of 2 cm and a height of 36 cm
D. a cylinder with a radius of 6 cm and a height of 4 cm

3. The radius of a cone is doubled while the height remains the same. The result is that the volume is multiplied by:

A. 2   B. 4
C. 6   D. 8

4. An ice cream log is packaged as a semi-circular cylinder as shown. What is the volume of the package in cubic inches?

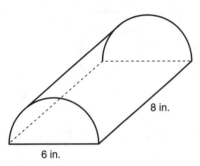

8 in.

6 in.

A. 18π B. 24π
C. 36π D. 72π

5. Find the number of cubic centimeters in the volume of the solid shown.

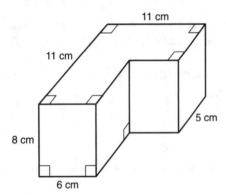

11 cm

11 cm

5 cm

8 cm

6 cm

6. Two cylinders have the same height. If one cylinder has a diameter that is three times the diameter of the other cylinder, how do the volumes of the two cylinders compare?

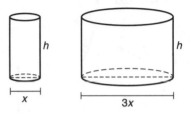

h h

x $3x$

7. A structure is formed with a cone attached to a cylinder. Find the volume of the structure. Show how you use formulas to find the answer. Express your answer to the nearest cubic inch.

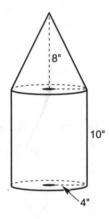

8"

10"

4"

8. Lindsey has 40 blocks, each one a cube with a volume of 1 cubic centimeter. List the dimensions of all the different rectangular prisms she can build using all the blocks for each prism. Explain why none of these prisms will turn out to be a cube.

2 C 4 Surface Area

The **surface area** of a solid is the sum of the areas of all the surfaces of the solid. Surface area is measured in square units.

To Find the Surface Area of a Solid:

1. Visualize the unfolded pattern (net) made up of the faces of the solid.
2. Find the area of each face of the solid.
3. Find the sum of the areas.

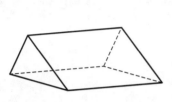
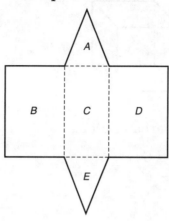

Surface area of the triangular prism = Area A + Area B + Area C + Area D + Area E

Model Problem

1. Find the surface area of the triangular prism shown.

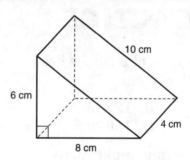

Solution   The surface area of the prism equals the area of two congruent triangular faces plus the area of three rectangular faces.

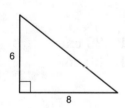
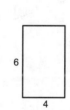
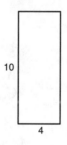
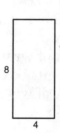

$A = \frac{1}{2}bh$ $A = lw$ $A = lw$ $A = lw$

$A = \frac{1}{2}(8)(6)$ $A = (6)(4)$ $A = (10)(4)$ $A = (8)(4)$

$A = 24$ $A = 24$ $A = 40$ $A = 32$

S.A. of prism = 2(24) + 24 + 40 + 32

Answer   144 cm^2

2. If the surface area of the cube is 24 cm², what is the length of an edge of the cube?

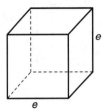

Solution   S.A. of cube = 6 × area of a face.

$$24 = 6 \cdot e^2$$
$$4 = e^2$$
$$\sqrt{4} = \sqrt{e^2}$$
$$2 = e$$

Answer   An edge is 2 cm.

 **PRACTICE**

1. Find the surface area of a cube whose edge has a length of 5 cm.

   A. 100 cm²   B. 120 cm²
   C. 125 cm²   D. 150 cm²

2. Which of the following applications would require finding surface area?

   A. finding the amount of packing material needed to fill a box
   B. finding the amount of ribbon needed to tie a bow on a gift box
   C. determining the amount of wrapping paper needed for a gift box
   D. determining the amount of water needed to fill a fish tank

3. The diagram shows a structure made with eight cubes. If each cube has an edge of 1 cm, what is the surface area of the structure?

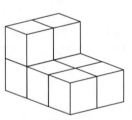

   A. 22 cm²   B. 26 cm²
   C. 28 cm²   D. 30 cm²

4. Which of the following solids has the greatest surface area?

A.

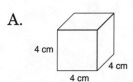

B.

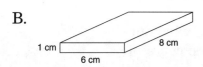

C.

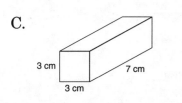

D.

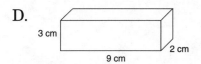

5. The volume of a cube is 27 cubic inches. Find the surface area. Show your procedure.

6. The bottom of a box measures 18 cm by 20 cm. The box is 10 cm high and has no top. What is the surface area of the box?

7. Two rectangular prisms each have a volume of 60 cubic inches. If each prism has a height of 10 inches, show that they do not have to have the same surface area. (Assume that all dimensions are whole numbers.) Use a diagram with your explanation.

Open-Ended Questions

8. Sketch the figure that the net shown folds into.

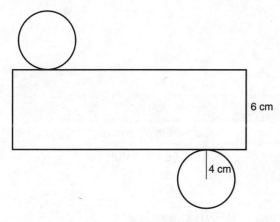

a. What is the solid called?
b. Find its volume. Show your work.
c. Find its surface area. Show your work.

9. a. A rectangular solid has a surface area of 42 square units. If the dimensions are each tripled, find the new surface area. Show your procedure.

b. What is the relationship between the original surface area and the surface area after the dimensions are tripled?

2 C 5 Standard and Non-Standard Units of Measure

Measurement Systems

The **metric system** of measurement is a decimal system. The basic unit for length is the **meter**, for weight is the **gram**, and for capacity is the **liter**.

The names of metric units are made by combining a prefix with one of the basic units (meter, gram, or liter). The prefixes name powers of ten, as shown in the following table.

Metric Prefixes						
1000	100	10	1	0.1	0.01	0.001
kilo-	hecto-	deka-	basic unit	deci-	centi-	milli-

Examples

a. 1 kilometer is a unit of length equal to 1000 meters.
b. 1 hectogram is a unit of weight equal to 100 grams.
c. 1 centiliter is a unit of capacity equal to 0.01 liter.

To convert from a larger unit to a smaller unit in the metric system, multiply the larger unit by the appropriate power of ten. To convert from a smaller unit to a larger unit, divide the smaller unit by the appropriate power of ten. To figure out what the appropriate power of ten is, count how many places you have to move on the metric prefix table.

Example   Convert 235 kilograms to decigrams.

Since this conversion is from a larger unit to a smaller unit, multiply. Since kilograms are 4 moves away from decigrams on the prefix table, multiply by 10^4, which is 10,000.

$$235 \times 10,000 = 2,350,000 \text{ decigrams}$$

Some common metric conversions are shown here.

Metric Conversions		
Length	**Weight (Mass)**	**Capacity (Liquid)**
1 meter (m) = 100 cm 1 kilometer (km) = 1000 m	1 gram (g) = 1000 mg 1 kilogram (kg) = 1000 g	1 liter (L) = 1000 ml

The following table shows the most commonly used units of measure in the **customary system** of measures.

To convert from a larger unit to a smaller unit in the customary system, multiply the larger unit by the appropriate conversion factor. To change from a smaller unit to a larger unit, divide the smaller unit by the appropriate conversion factor.

Customary Conversions		
Length	**Weight (Mass)**	**Capacity (Liquid)**
1 foot (ft) = 12 inches (in.)	1 pound (lb) = 16 ounces (oz)	1 cup (c) = 8 fluid ounces (fl oz)
1 yard (yd) = 36 inches	1 ton (T) = 2,000 pounds	1 pint (pt) = 2 cups
1 yard = 3 feet		1 quart (qt) = 2 pints
1 mile (mi) = 5,280 feet		1 gallon (gal) = 4 quarts

Example   Convert 5 yards to inches.

First convert from yards to feet:

$$5 \text{ yards} \times \frac{3 \text{ feet}}{\text{yard}} = 15 \text{ feet}.$$

Next convert from feet to inches:

$$15 \text{ feet} \times \frac{12 \text{ inches}}{\text{foot}} = 180 \text{ inches}.$$

There are 180 inches in 5 yards.

 Model Problem

1. A rectangle has a length of 85 cm and a width of 53 cm. What is the perimeter of the rectangle in meters?

Solution

$$P = 2(l + w)$$
$$P = 2(85 \text{ cm} + 53 \text{ cm})$$
$$= 2(138 \text{ cm}) = 276 \text{ cm}$$

To convert to meters, divide by 100 since 1 cm = 0.01 meter.

Answer $P = 2.76$ m

2. Dora purchased two packages of mix for veggie burgers. One package weighs 2.5 pounds and the other package weighs 3.75 pounds. How many four-ounce veggie burgers can Dora make from the two packages?

Solution   Total number of pounds purchased = 2.5 lb + 3.75 lb = 6.25 lb.

Since the veggie burgers are 4 ounces each, convert from pounds to ounces.

$$1 \text{ lb} = 16 \text{ oz}$$
$$6.25 \text{ lb} \times 16 \text{ oz per lb} = 100 \text{ oz}$$
$$100 \text{ oz} \div 4 \text{ oz} = 25$$

Answer   Dora can make 25 veggie burgers.

Estimating Measurements

In working with measurements, you can better judge the reasonableness of results if you know how common units of measure relate to familiar objects.

Length
An inch is about the distance between the joints of your index finger.
A foot is about the length of a sheet of notebook paper.
A yard is about the distance from the tip of your nose to the tip of your middle finger with your arm outstretched.
A meter is a little more than a yard.
A centimeter is a little less than half an inch.

Weight (Mass)
A gram is about the weight of a paper clip.
A kilogram is about the weight of a hammer.
An ounce is about the weight of a slice of bread.
A pound is about the weight of a loaf of bread.

 Model Problem

Which of the following is the best estimate for the measure of the diameter of an NBA basketball?

A. 4 in.
B. 8 in.
C. 13 in.
D. 24 in.

Solution   4 inches and 8 inches would be too small. 24 inches would be much too large.

Answer   13 inches is a reasonable estimate.

Accuracy of Measurements

In measuring length using an inch ruler, it is necessary to determine the degree of precision required.

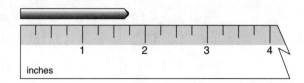

The chalk is 2 inches long to the nearest inch.

The chalk is $1\frac{1}{2}$ inches long to the nearest half inch.

The chalk is $1\frac{3}{4}$ inches long to the nearest quarter inch.

Model Problem

Find the length of the straw to the nearest half inch.

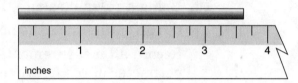

Solution   The length of the straw is between 3 and 4 inches. In finding the length to the nearest half inch, you need to see if it is closest to 3, $3\frac{1}{2}$, or 4. Since it is closest to $3\frac{1}{2}$, the length is $3\frac{1}{2}$ inches.

PRACTICE

1. A tree grows 1.4 cm each day. In 120 days, how many METERS will the tree have grown?

   A. 0.168 m   B. 1.68 m
   C. 16.8 m   D. 168 m

2. Which of the following is a correct measurement of the length of the crayon?

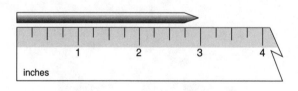

   A. 2 inches to the nearest inch

   B. $2\frac{1}{2}$ inches to the nearest half inch

   C. $2\frac{1}{2}$ inches to the nearest quarter inch

   D. 3 inches to the nearest half inch

3. If line segment $\overline{AB}$ is 3.5m long, about how long is line segment $\overline{AC}$?

A. 4 m   B. 10 m
C. 14 m   D. 18 m

4. Which of the following is a correct reading of the voltage?

Voltage

A. 4 volts to the nearest volt
B. 3.5 volts to the nearest volt
C. 3.5 volts to the nearest $\frac{1}{2}$ volt
D. 3.5 volts to the nearest $\frac{1}{4}$ volt

5. Each side of a regular decagon (10-sided figure) has a length of 4.25 inches. Find the perimeter.

A. 3 ft 6.5 in.   B. 3 ft 8 in.
C. 3 ft 11 in.   D. 4 ft

6. Approximately what percent of a yard is an inch?

A. 3%   B. 10%
C. 33%   D. 36%

7. At 6:00 A.M, the temperature in Nome, Alaska, was 15° below zero Fahrenheit. The average increase in temperature per hour was 3°. What was the temperature at noon?

A. −12°F   B. −3°F
C. 3°F   D. 33°F

8. 180 cm is a good estimate of

A. the length of a driveway
B. the length of a dining room table
C. the height of a kindergarten child
D. the length of a city block

9. 6 pounds is a good estimate of the weight of

A. a box of cereal
B. three apples
C. two math textbooks
D. a case of 24 packages of copy paper

10. Using a ruler, measure the sides of the given figure to the nearest $\frac{1}{8}$ in. Record all measurements. What is the perimeter of the figure to the nearest eighth of an inch?

11. How many times does a person's heart beat in one day if it beats an average of 68 times per minute?

12. Temperature can be measured in either of two systems:
Celsius (0° freezing point of water; 100° boiling point of water) and *Fahrenheit* (32° freezing point of water; 212° boiling point of water). The two systems are related by the formula:

$$F = \frac{9}{5}C + 32°.$$

20°C is considered an appropriate measure for room temperature. Express this measure in degrees Fahrenheit.

13. Use an inch ruler to determine the lengths of the sides of the figure shown. Use your measurements to find the area of the figure.

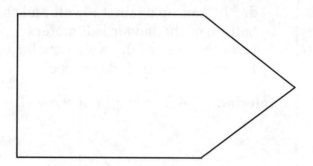

---

2 C 6 Pythagorean Theorem

If you know the lengths of two sides of a right triangle, a special formula can be used to find the length of the third side.

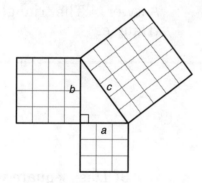

The number of square units in the square on the hypotenuse, c, is equal to the sum of the number of square units in the squares on both legs, a and b.

$$\boxed{c^2 = a^2 + b^2}$$

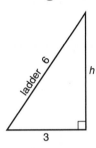

Model Problem

1. A ladder is 6 meters long. If the ladder is leaning against a wall and the bottom of the ladder is 3 meters from the base of the wall, how far up the wall does the ladder reach?

Solution   Let h = height of the wall.

$$c^2 = a^2 + b^2$$
$$6^2 = 3^2 + h^2$$
$$36 = 9 + h^2$$
$$27 = h^2$$
$$\sqrt{27} = \sqrt{h^2}$$
$$5.2 \approx h$$

Answer   The ladder reaches about 5.2 meters up the wall.

2. Is a triangle with sides of lengths 5, 9, and 10 a right triangle?

Solution   If the triangle is a right triangle, the lengths of the sides must satisfy the Pythagorean Theorem. Conversely, if the lengths of the three sides satisfy the Pythagorean Theorem, the triangle must be a right triangle. Put the side lengths into the formula to check.

$$5^2 + 9^2 \stackrel{?}{=} 10^2$$
$$25 + 81 \stackrel{?}{=} 100$$
$$106 \neq 100$$

Answer   The triangle is NOT a right triangle.

PRACTICE

1. Which of the following represents the lengths of the sides of a right triangle?

A. 2, 3, 4   B. 3, 4, 6
C. 6, 8, 10   D. 4, 6, 8

2. A rectangle has dimensions 5 cm by 12 cm. How many centimeters long is a diagonal of the rectangle?

A. 12   B. 13   C. 17   D. 30

3. The two legs of a right triangle have lengths of 5 and 6 units. Between what two integers would the length of the hypotenuse fall?

A. 5 and 6   B. 6 and 7
C. 7 and 8   D. 8 and 9

4. *ABCD* is a square with a perimeter of 40 inches. What is the perimeter of the shaded figure?

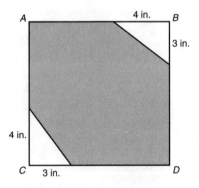

A. 28 inches   B. 36 inches
C. 40 inches   D. 88 inches

5. On Saturday, Winn and Pete left their cabin and hiked 4 km east and then 7 km north to a bubbling brook, where they ate snacks. Which of the following is the most reasonable answer for the distance between the bubbling brook and the cabin?

A. 6.5 km   B. 8.1 km
C. 9 km   D. 11 km

6. An 8-foot ladder leaning against a wall reaches 6 feet up the wall. How far from the base of the wall is the bottom of the ladder?

A. 2 ft   B. 5.3 ft
C. 6.2 ft   D. 10 ft

7. Find the value of b in this right triangle.

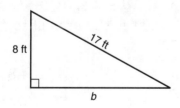

8. Each diagonal of a rectangle has a length of 10 yards. If one dimension of the rectangle is 6 yards, find the perimeter.

9. *ABCD* represents a rectangular piece of land with dimensions 5 feet by 22 feet. $\overline{EC}$ is a straight path from E to C. How many feet do you save if you go from A to C by taking $\overline{EC}$ instead of A to C by way of B? Explain your answer.

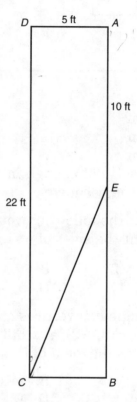

10. Triangle *ABC* is isosceles. Show how you can use the Pythagorean Theorem to find the height of the triangle. Give the steps used in your thinking.

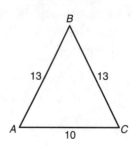

1. What is the perimeter of the figure shown?

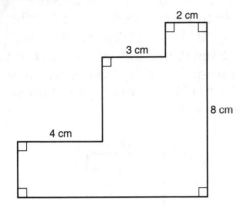

A. 17 cm
B. 34 cm
C. 72 cm
D. There is not enough information to find the perimeter.

2. Which of the following would NOT represent the sides of a right triangle?

A. 1, 1, $\sqrt{2}$ B. $\sqrt{3}$, $\sqrt{2}$, $\sqrt{5}$
C. 8, 15, 17   D. 5, 6, $\sqrt{11}$

3. A rectangle has vertices at $P(3, 10)$, $Q(8, 10)$, $R(8, -2)$, and $S(3, -2)$. How long is the diagonal of this rectangle?

A. 12   B. $\sqrt{150}$
C. 12.5   D. 13

4. The dimensions of rectangular Pigeon Park are 66 yards by 88 yards. Lesley needs to walk from the southeast corner of the park to the northwest corner. How many yards longer is it to walk along the edges than to walk along the diagonal?

A. 4   B. 44
C. 54   D. 110

5. The given figure consists of five congruent squares. If the total area is 80 square units, what is the perimeter of the figure?

A. 12   B. 24
C. 36   D. 48

6. Which of the following is NOT true?

A. 20 ft < 7 yd
B. 100 in. < 3 yd
C. 198 cm < 2 m
D. 2010 m < 2 km

7. In which list below are the units of measurement for area arranged in order from LEAST to GREATEST?

A. in.^{2}, cm^2, km^2
B. km^2, m^2, cm^2
C. cm^2, m^2, km^2
D. cm^2, km^2, in.^{2}

8. What is the LEAST number of circles with a radius of 1 inch that give a total area greater than 30 square inches?

A. 9   B. 10
C. 15   D. 30

9. Rectangle $ABCD$ is similar to rectangle $WXYZ$, with $\overline{AB}$ corresponding to $\overline{WX}$. If $\overline{AB}$ = 24, $\overline{BC}$ = 30, and $\overline{WX}$ = 16, what is the area of rectangle $WXYZ$?

A. 20   B. 204.8
C. 320   D. 720

10. A circle has a diameter with endpoints at $(-4, 0)$ and $(4, 0)$. What is the circumference of the circle?

A. 4π B. 8π
C. 16π D. 64π

11. Find the volume of the three-dimensional object formed from the following net.

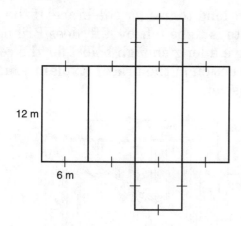

12 m

6 m

  A. 360 m³
  B. 432 m³
  C. 864 m³
  D. There is not enough information to find the volume.

12. Find the area of the rectangle if the radius of each circle is 3 cm.

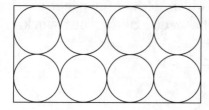

  A. 48 cm²   B. 72 cm²
  C. 144 cm²   D. 288 cm²

13. Which of the following statements is TRUE concerning the capacities of the two containers?

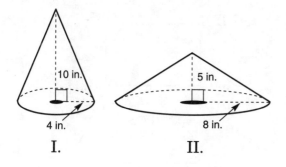

10 in.   5 in.

4 in.   8 in.

I.   II.

  A. The capacity of I is greater than the capacity of II.

  B. The capacity of II is greater than the capacity of I.
  C. The capacity of II is the same as the capacity of I.
  D. Based on the given information, it is not possible to determine the comparison between the capacities.

14. The diagram consists of a rectangle and two semicircles. If the rectangle has dimensions of 4 by 10, what is the total perimeter of the figure, to the nearest tenth of a unit?

15. An isosceles trapezoid has vertices at (2, 0), (10, 0), (7, 4), (5, 4). Find the perimeter of the trapezoid.

16. How many circles 1 inch in diameter would it take to have a combined area equal to the area of a single circle 4 inches in diameter?

17. If the volume of the prism is 86.4 cubic centimeters, find the missing dimension.

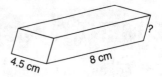

4.5 cm   8 cm

18. A can of paint covers 750 square feet. A parking lot contains 60 cylindrical poles of the dimensions shown. How many cans of paint are needed to give each pole *two* coats of paint?

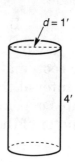

d = 1'

4'

Open-Ended Questions

19. How will the surface area of a cylinder change if the height is doubled and the radius is cut in half? Explain your thinking and use a diagram.

20. A baseball diamond is a square with 90-foot sides. The pitcher's mound is exactly 66′ 6″ from home plate. How far is the pitcher's mound from second base? Show your procedure in doing the problem.

21. The solid has faces that are right triangles and rectangles. What is the total surface area of the solid? Explain your procedure.

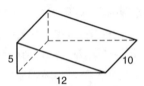

22. A diagonal brace on a wooden gate would look as pictured. In building the gate, Briana has a board that is 7 ft long to use for the brace. If the gate is to be 6 ft by 4 ft, does Briana have a long enough board for the construction of the brace? Explain your answer.

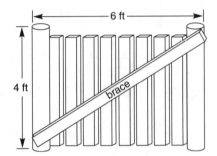

23. Malcolm has a circular table with a diameter of 4 feet. He plans to cover the table with a circular tablecloth that has a one foot overhang all around the table. He wants to sew fringe around the entire edge of the tablecloth. How many feet of fringe will he need? Show your work.

ASSESSMENT Cluster 2

1. How many edges meet at each vertex of a cube?

A. 2   B. 3   C. 4   D. 6

2. Two vertices of a triangle are (0, 0) and (6, 0). Which of the following points would NOT give you a right scalene triangle?

A. (0, −5)   B. (6, 4)
C. (−3, 0)   D. (6, −4)

3. Point A with coordinates (8, 15) is on a circle with center at the origin. Find the coordinates of point B such that $\overline{AB}$ is a diameter of the circle.

A. (−15, −8)   B. (−8, −15)
C. (8, −15)   D. (4, 7.5)

4. Which of the following figures show a line of symmetry for the figure?

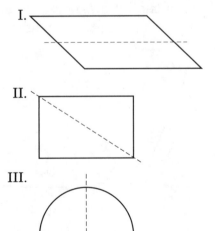

A. II only   B. III only
C. II and III   D. I, II, and III

5. A triangle has vertices at A (−3, 1), B(3, 1), and C(0, −5). How many lines of symmetry does the triangle have?

   A. 0   B. 1   C. 2   D. 3

6. If you spin a certain two-dimensional figure about a vertical axis, a cylinder results. What type of two-dimensional figure did you start with?

   A. right triangle   B. semicircle
   C. rectangle   D. trapezoid

7. How many different isosceles triangles can you find, with sides that are whole numbers, such that the perimeter of the triangle is 20?

   A. 3   B. 4   C. 5   D. 6

8. A rectangular swimming pool is 72 feet long and 32.5 feet wide. The pool is surrounded by a concrete walkway that is 3.5 feet wide. What is the area of the walkway?

   A. 227.5 square feet
   B. 237 square feet
   C. 378 square feet
   D. 780.5 square feet

9. If the circumference of a circle is divided by the length of a radius, the quotient is:

   A. $\frac{1}{2}\pi$ B. π
   C. 2π D. 2

10. Find the height of a trapezoid with an area of 42.5 square centimeters if one base is 5.2 cm and the other base is 3.3 cm.

    A. 4.95 cm   B. 5 cm
    C. 10 cm   D. 38.25 cm

11. Which of the following times would NOT represent an obtuse angle made by the hands of a clock?

    A. 5:05   B. 6:59
    C. 12:15   D. 12:25

12. If 1 in. ≈ 2.5 cm, then the length of a diagonal of a sheet of paper measuring $8\frac{1}{2}$ in. by 11 in. is closest to:

    A. 35 cm   B. 35 mm
    C. 70 cm   D. 70 mm

13. Anytown, USA, has a cylindrical water tank with dimensions as shown. As a result of increased demand for water, the council of Anytown wants to build a new cylindrical tank with twice the volume of the original tank. Which of the following options could be used?

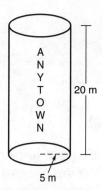

    I. Double the height of the tank and maintain the same diameter.
    II. Double the diameter of the tank and maintain the same height.
    III. Double both the diameter and the height of the tank.

    A. I only   B. II only
    C. III only   D. I and II

14. Which list shows the three figures *P*, *Q*, and *R* in order of size, from LEAST to GREATEST area?

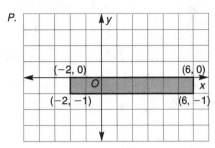

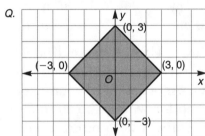

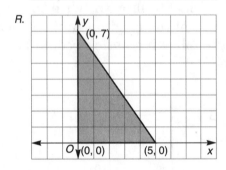

A. *P, Q, R*   B. *Q, P, R*
C. *R, P, Q*   D. *P, R, Q*

15. The Bagel Express uses an average of 2 lb 8 oz of cream cheese per hour. If the Bagel Express is open 7 hours a day for 7 days per week, about how much cream cheese would be used in 3 weeks?

A. 105 lb   B. 122.5 lb
C. 294 lb   D. 367.5 lb

16. Which of the following is NOT possible for any type of triangle regarding the number of lines of symmetry?

A. A triangle can have no lines of symmetry.
B. A triangle can have exactly one line of symmetry.
C. A triangle can have exactly two lines of symmetry.
D. A triangle can have three lines of symmetry.

17. Which of the following rectangles are similar?

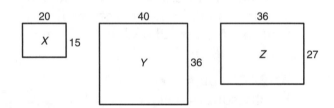

A. *X* and *Y* only
B. *X* and *Z* only
C. All three rectangles are similar.
D. None of the rectangles are similar.

18. The points (2, 5) and (2, −5) are the endpoints of a diameter of a circle. Find the area of the circle.

A. 4π square units
B. 5π square units
C. 25π square units
D. 100π square units

19. Which of the following sets of coordinates could represent the vertices of an isosceles trapezoid?

A. {(0, 0), (6, 0), (7, 4), (0, 4)}
B. {(0, 0), (6, 0), (6, 4), (0, 4)}
C. {(−4, 0), (0, 4), (4, 0), (0, −4)}
D. {(−4, 0), (4, 0), (2, 8), (−2, 8)}

20. If the letter N is reflected over the *y*-axis, which of the following represents the image?

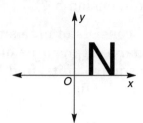

A.

B.

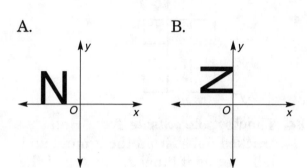

C.

D.

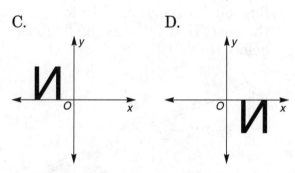

21. If the measure of ∠*ECF* is 15°, what is the measure of ∠*ACD*?

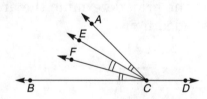

A. 45°   B. 135°
C. 145°   D. 165°

22. *ABCD* is a square. If you remove a small right triangle at vertices *A* and *C* (with right angles at *A* and *C*), what type of figure remains?

A. octagon   B. hexagon
C. pentagon   D. square

23. Draw a three-dimensional figure that has all three of the following characteristics:

- an odd number of faces
- an even number of vertices
- more vertices than faces

24. *ABCD* is a square and the two triangles are equilateral. What is the measure of ∠*EBF*?

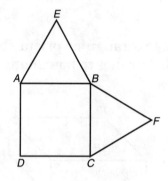

25. Triangle *ABC* has vertices at *A*(−4, 2), *B*(−4, 8), and *C*(−2, 2). What are the coordinates of the image of the triangle after it is translated 6 units to the right and 2 units down?

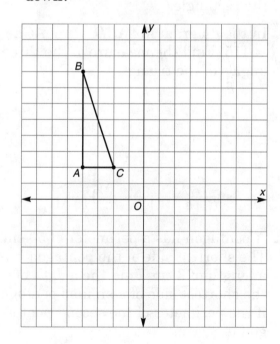

26. How many revolutions will it take a 24-inch-diameter bicycle wheel to travel 1 mile (5,280 feet)? Give your answer to the nearest whole number.

27. Lines l and m are parallel. Find m$\angle x$.

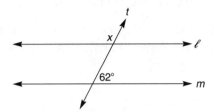

28. In the rectangular prism shown, edge $\overline{BC}$ is parallel to how many other edges?

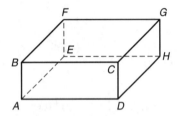

29. A piece of wire 30 cm long has been bent to form a rectangle. The area enclosed by the rectangle is 50 square centimeters. Find the longer side of the rectangle.

30. In the diagram m$\angle ABD$ = m$\angle ACE$ = 115. Find the measure of the three angles of the triangle.

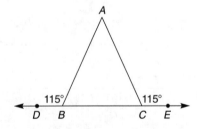

31. A pentagon has a perimeter of 50 cm. The shortest side of the pentagon has a length of 3 cm. Find the length of the shortest side of a similar pentagon if the perimeter of the second pentagon is 80 cm.

32. A rectangular solid has a volume of 227.5 cubic inches. If two dimensions are 5 inches and 7 inches, what would the third dimension be?

33. The figure consists of nine small congruent squares. If the area of the figure is 900 square units, find the perimeter of the figure.

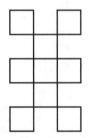

34. The flagpole outside Town Hall cracked 5 feet from the ground and fell over as if hinged. The top of the flagpole hit the ground 12 feet from the base. How tall was the flagpole before it fell?

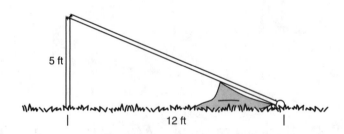

35. Using the grid, determine the area of the shaded figure.

Open-Ended Questions

36. Find the area and perimeter of the trapezoid. Show your complete procedure.

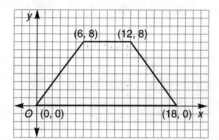

37. Explain why the perimeter of the figure must be greater than 20 units. Also, show how to obtain a good approximation for the perimeter.

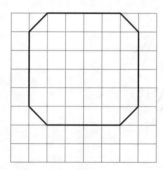

38. a. Explain why a triangle with sides 5, 10, and 12 cannot be a right triangle.
   b. Write another set of lengths of sides of a triangle that could NOT be the sides of a right triangle.
   c. Write two sets of lengths that could be the sides of a right triangle.

39. In the grid, there is a figure with a certain area.

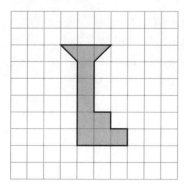

a. On graph paper, draw a rectangle with an area equal to the area of the figure in the grid.
b. On graph paper, draw a triangle with an area equal to the area of the figure in the grid.

40. a. Plot the points (8, 6), (8, 2), and (14, 2) on graph paper. Connect the points to form a triangle. What type of triangle is formed?
   b. Draw the triangle you get when you dilate the triangle in part a by a scale factor of 0.5. How are the two triangles related?
   c. The area of the smaller triangle is what percent of the area of the larger triangle? Explain why.

41. Use a set of tangrams to build the rectangle pictured below.

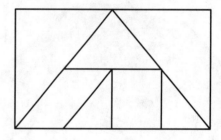

a. Move one piece of the rectangle to change the figure into a parallelogram with no right angles. Sketch your figure.
b. Write a paragraph to explain how the area and perimeter of the original rectangle compare to the area and perimeter of the new parallelogram. Be sure to explain how you know this relationship exists. It is not necessary to actually calculate the area and perimeter of the figures.

CUMULATIVE ASSESSMENT

Clusters 1 and 2

1. Which of the following is NOT correctly expressed in scientific notation?

   A. 6.14×10^3 B. 8×10^{-3}
   C. 56.2×10^4 D. 6.1×10^{-5}

2. The following numbers represent volumes of a cube. For which value would the cube NOT have an edge that is a whole number?

   A. 125 B. 300
   C. 1,000 D. 3,375

3. A Hudson University sweatshirt has a wholesale price of $30.00. The Sports Exchange marks up the wholesale price 20% to get the retail price. During a sale, the same sweatshirt is priced at 15% off the retail price. What is the sale price?

   A. $20.40 B. $30.60
   C. $31.50 D. $36.00

4. If the area of a rectangle is 10 square units, what is the ratio of the length to the width?

   A. 1 to 10
   B. 2 to 5

   C. 5 to 2
   D. It cannot be determined from the information given.

5. How many two-digit numbers have all of the following characteristics?

   • The number is a multiple of 16.
   • The number is a factor of 144.
   • The number is divisible by 3.

   A. 0 B. 1
   C. 2 D. 3

6. How many of the eight angles in the diagram need to be measured with a protractor in order to figure out the number of degrees in the rest?

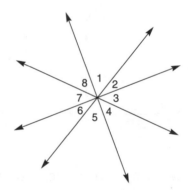

   A. 1 B. 2 C. 3 D. 4

7. A line segment has endpoints at (1, 6) and (4, 6). The segment is translated 4 units to the right and 3 units down, and then reflected over the x-axis. After the reflection, what are the coordinates of the endpoints of the final segment?

   A. (5, 3) and (8, 3)
   B. (−5, 3) and (−8, 3)
   C. (4, −2) and (7, −2)
   D. (5, −3) and (8, −3)

8. On a map, 1 centimeter represents 125 kilometers. How many kilometers apart are two cities that are 48 millimeters apart on the map?

A. 60 km   B. 600 km
C. 6000 km   D. 60,000 km

9. Which of the following changes does NOT show a 10% increase?

A. $100 \rightarrow 110$ B. $50 \rightarrow 60$
C. $10 \rightarrow 11$ D. $\dfrac{1}{10} \rightarrow \dfrac{11}{100}$

10. If you double the radius of the base of a cylinder and also double the height of the cylinder, what percent increase will there be in the volume of the cylinder?

A. 200%   B. 400%
C. 700%   D. 800%

11. Using the data from the table, which of the following correctly lists the surface areas of the three boxes from LEAST to GREATEST?

Box	Dimensions (in centimeters)
P	$6 \times 6 \times 6$
Q	$4 \times 10 \times 2$
R	$5 \times 8 \times 3$

A. P, Q, R B. R, Q, P
C. Q, R, P D. Q, P, R

12. $\triangle ABC$ and $\triangle ACD$ are right triangles. Determine the perimeter of quadrilateral $ABCD$. Give your answer to the nearest tenth of a unit.

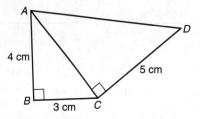

13. Evaluate: $28 \div 4 - 2(8 - 7)^2 + 5 \times 3$.

14. Sketch two additional views of the figure shown from different perspectives.

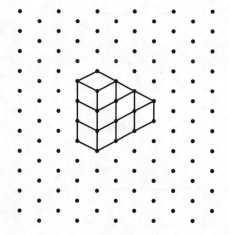

15. The lengths of the sides of a triangle are 36, 30, and 18. If the longest side of a similar triangle measures 9 units, what is the length of the shortest side of this triangle?

Open-Ended Questions

16. As a back-to-school incentive on Wednesdays, Supply Shack gave every fifteenth customer a free pen and every twenty-fifth customer a free notebook. On a particular Wednesday, Supply Shack had 300 customers.

a. How many free pens were given away on that Wednesday?

b. How many free notebooks were given away on that Wednesday?

c. Did any customers receive a free pen and a free notebook? If so, how many customers?

d. If pens sell for 79¢ and notebooks sell for $1.19, how much did the Supply Warehouse lose in income by giving away these items?

17. Consider the two triangles shown.

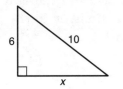 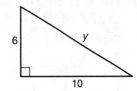

a. To the nearest tenth of a unit, find the value of x and the value of y.

b. Explain why the two triangles are not similar.

18. Some shapes have one or more lines of symmetry and others have none.

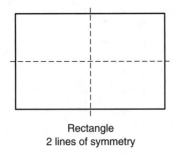

Rectangle
2 lines of symmetry

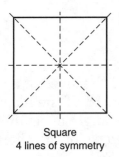

Square
4 lines of symmetry

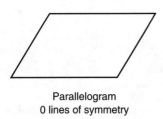

Parallelogram
0 lines of symmetry

a. Draw a triangle with exactly one line of symmetry. Show the line of symmetry.

b. Draw a triangle with three lines of symmetry. Show the lines of symmetry.

c. Draw a triangle with no lines of symmetry.

d. Show how you can join a square and two triangles so that the resulting figure has two lines of symmetry.

19. The formula for the volume of a cylinder is $V = \pi r^2 h$. If the height of a cylinder is 10 cm, what is the smallest whole number of centimeters for the length of the radius that would produce a volume of at least 500 cubic centimeters? Explain your procedure.

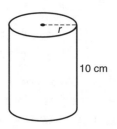

10 cm

20. 3, 5, and 8 are factors of x.

a. Could x be a whole number less than 100? Explain

b. Explain why x could not be an odd number.

c. List four other numbers that would also be factors of x.

21. A square on a coordinate graph has horizontal and vertical sides. The upper left vertex of the square is located at $(-3, 4)$. The perimeter of the square is 32 units.

a. Draw a sketch of the square and indicate the coordinates of the other three vertices.

b. What percent of the area of the square is in each quadrant?

Extra Practice

Open-Ended Questions

Study Questions 1 and 2, which are presented with their solutions, and then try Questions 3 and 4.

When responding to an open-ended question, think about what you must do to form a response that will receive a score of 3.

- Answer all parts of the question.
- Present your work clearly, so that the person grading it will understand your thinking.
- Show all your work, including calculations, diagrams, and written explanations.

1. Rectangle $ABCD$ has vertices in the first quadrant: $A(2, 3)$, $B(2, 7)$, $C(8, 7)$, and $D(8, 3)$.

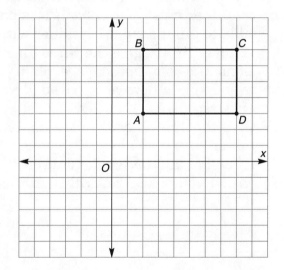

- Give the coordinates of the vertices of any rectangle in the second quadrant that is congruent to rectangle $ABCD$.
- What is the perimeter of rectangle $ABCD$?
- What is the area of rectangle $ABCD$?

Solution (for a score of 3)

Second quadrant: $(-1, 3)$, $(-1, 9)$, $(-5, 9)$, $(-5, 3)$

Perimeter of $ABCD$ = 2(length + width)

$$P_{ABCD} = 2(6 + 4)$$

$$P_{ABCD} = 2(10) = 20 \text{ units}$$

Area of $ABCD$ = length $\times$ width

$$A_{ABCD} = 6 \times 4$$

$$A_{ABCD} = 24 \text{ square units}$$

2. A rectangular prism is shown below.

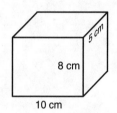

- Find the volume of the figure.
- If you double one dimension, triple another dimension, and leave the third dimension alone, what does this do to the volume? Express your answer as a ratio. Show how you arrived at your answer.
- Make a table showing four different sets of values for the dimensions of rectangular prisms having the same volume as the one shown.

Solution (for a score of 3)

$$V = lwh$$
$$V = 10(8)(5) = 400 \text{ cm}^3$$

8 cm $\rightarrow$ triple $\rightarrow$ 24 cm
10 cm $\rightarrow$ double $\rightarrow$ 20 cm
5 cm $\rightarrow$ same $\rightarrow$ 5 cm
$$V = (24)(20)(5) = 2400 \text{ cm}^3$$

The ratio of the original volume to the changed volume is 400:2400, which simplifies to 1:6.

$$\frac{400}{2,400} = \frac{4}{24} = \frac{1}{6}$$

The volume has been multiplied by 6.

The table shows different sets of values for the dimensions of rectangular prisms that all have a volume of 400 cm^3.

l	w	h	V
20	10	2	400
5	5	16	400
10	10	4	400
40	10	1	400

3. Isosceles triangle ABC has vertices in the first quadrant: $A(1, 1)$, $B(7, 1)$, and $C(4, 5)$.

   • Plot $\triangle ABC$ on graph paper. Label the vertices.
   • Find the perimeter of $\triangle ABC$. Show your work.
   • What are the coordinates of the image of $\triangle ABC$ after a translation 4 units left and 3 units down? Plot the image, $\triangle A'B'C'$ on the coordinate graph.

4. An industrial designer is hired to construct a storage container with a volume of 216 ft^3. The customer says that the container must be in the shape of a rectangular prism and that all dimensions must be whole numbers.

   • Give the dimensions (length, width, and height) of two rectangular prisms that would meet the customer's needs.
   • The customer later decides that the container must be a cube. If it is possible to build a cubic container with a volume 216 ft^3, give the three dimensions. If it is not possible, explain why. Explain how you arrived at your answer.
   • Give the dimensions of one cubic container with a volume LESS than 216 ft^3 and of another cubic container with a volume GREATER than 216 ft^3.

CLUSTER 3

Data Analysis, Probability, Statistics, and Discrete Mathematics

MACRO A

Predict, determine, interpret, and use probabilities.

3 A 1 Probability of Simple Events

Probability is used to describe how likely it is that a particular event will occur. A **sample space** is the set of all the possible outcomes for some activity or experiment. The **probability** of an event is defined as the ratio of the number of favorable outcomes to the total number of possible outcomes.

$$\text{probability} = \frac{\text{number of favorable outcomes}}{\text{total number of possible outcomes}}$$

The probability of an event can be expressed as a fraction, decimal, or percent with a value greater than or equal to zero and less than or equal to one ($0 \le P \le 1$).

If $P(A) = 0$, then it is impossible for A to occur.
If $P(A) = 1$, then it is certain the event will occur.

When two or more possible outcomes of a given situation have the same probability, the outcomes are considered *equally likely*.

A game in which each player has an equally likely chance of winning is considered a *fair game*.

Example Jay will flip a coin to decide whether to go to the movies. If he gets heads, he will go to the movies.

a. The *activity* or *experiment* is flipping a coin.
b. The *sample space* is {heads, tails}.
c. The *favorable outcome* is heads. The *possible outcomes* are heads and tails.
d. $P(\text{heads}) = \frac{1}{2}$. This can also be expressed as $P(\text{heads}) = .5$ and $P(\text{heads}) = 50\%$.

 # Model Problem

1. Helen rolls a standard die. Find the probability of each of the following events.

   a. rolling a 5
   b. rolling an even number
   c. rolling a number less than 3
   d. rolling a number greater than 7
   e. rolling a number less than 7

Solution The sample space consists of {1, 2, 3, 4, 5, 6}, so there are 6 possible outcomes. The probabilities can be determined by applying the definition of probability.

a. $P(5) = \frac{1}{6}$ (since 5 is the only favorable outcome)

b. $P(\text{even}) = \frac{3}{6}$ (since 2, 4, and 6 are favorable outcomes)

c. $P(< 3) = \frac{2}{6}$ (since 1 and 2 are favorable outcomes)

d. $P(> 7) = 0$ (since there are no favorable outcomes)

e. $P(< 7) = \frac{6}{6} = 1$ (since all outcomes are favorable)

Note: Answers to some probability questions may appear as simplified fractions. For example, the answer to part b might appear as $\frac{1}{2}$.

2. Troy must select one number randomly from the set of numbers from 21 to 30. Find the probability that he will select each of the following outcomes. Express each probability as a fraction, a decimal, and a percent.

   a. an even number
   b. a perfect square
   c. a multiple of 3
   d. a number greater than 15

Solution The sample space consists of 10 numbers: {21, 22, 23, 24, 25, 26, 27, 28, 29, 30}. Use the definition of probability to write the probability fractions. Since the denominator in each case will be 10, it will be easy to convert the fractions to decimals and percents.

a. $P(\text{even}) = \frac{5}{10} = \frac{1}{2}$ or .5 or 50%

b. $P(\text{perfect square}) = \frac{1}{10}$ (since 25 is the only favorable outcome) = .1 = 10%

c. $P(\text{multiple of 3}) = \frac{4}{10}$ (since 21, 24, 27, and 30 are favorable outcomes) $\frac{4}{10}$ simplifies to $\frac{2}{5}$, and $\frac{4}{10} = .4 = 40\%$

d. $P(\text{number} > 15) = \frac{10}{10} = 1 = 100\%$ (since all outcomes are favorable)

Experimental and Theoretical Probabilities

Experimental probability results from conducting an experiment or making an observation many times and recording the results. For example, if you toss a coin 50 times and obtain heads 30 times, the experimental probability of obtaining heads would be $\frac{30}{50}$ or $\frac{3}{5}$. Each time you toss the coin 50 times, the experimental probability of heads may vary.

In contrast with experimental probability, **theoretical probability** represents what you would expect from the *theory* or description of the situation. When we say that the probability of obtaining heads on the toss of a coin is $\frac{1}{2}$ or 50%, or that the probability of obtaining an ace from deck of cards is $\frac{4}{52}$ or $\frac{1}{13}$, we are giving the theoretical probability.

 Model Problem

1. The table shows the results of data gathered on the number of girls in 100 families with four children. For example, 10 families had no girls.

   a. Based on the data, what is the experimental probability that exactly two children will be girls?

Girls in Families of Four Children	
Number of Girls	Frequency
0	10
1	15
2	49
3	19
4	7

   b. Explain why your answer would most likely be different if you collected data from another group of 100 families consisting of four children.

Solution

a. From the information in the table, the experimental probability is $\frac{49}{100}$.

b. The probabilities are based on the frequencies. As the data are gathered, it is not likely that you would get the same frequencies.

2. What is the theoretical probability of having exactly two girls in a family of four children?

BBBB	BGBB	GGGG	GBGG
BBBG	BGBG	GGGB	GBGB
BBGB	BGGB	GGBG	GBBG
BBGG	BGGG	GGBB	GBBB

Solution The sample space shows a total of 16 possibilities for combinations of boys and girls in a family with four children.

Since 6 of the sequences above represent exactly two girls (and two boys), the theoretical probability is $\frac{6}{16}$ or $\frac{3}{8}$ or 37.5%.

PRACTICE

1. Which of the following cannot be the answer to a probability question?

A. 0   B. 30%

C. $\frac{11}{10}$ D. $\frac{10}{11}$

2. You flip a fair coin. The first eight flips come up heads. What is the probability that the ninth flip of the coin will be a tail?

A. $\frac{1}{2}$ B. 1

C. $\frac{8}{9}$ D. $\frac{1}{9}$

3. Consider the following events:
   I. Obtaining a sum of 2 in rolling two dice
   II. Obtaining 5 heads when tossing five coins
   III. Obtaining a red with one spin of the spinner shown

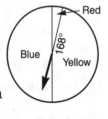

Arrange the events in order from LEAST probable to MOST probable.

A. III, II, I   B. I, II, III
C. I, III, II   D. II, I, III

4. Two events, A and B, are considered *complementary* if

$$P(B) = 1 - P(A).$$

For example, in tossing a coin, P(heads) $= \frac{1}{2}$ and P(tails) $= \frac{1}{2}$. Since $\frac{1}{2} = 1 - \frac{1}{2}$, the events of tossing heads and tossing tails are complementary. Which of the following would NOT represent complementary events?

A. spinning an odd number on the spinner shown; spinning an even number on the spinner shown

B. picking a number divisible by 3 from the set of whole numbers between 1 and 30; picking a number not divisible by 3 from the set of whole numbers between 1 and 30.

C. obtaining a sum less than 7 when tossing two dice; obtaining a sum greater than 7 when tossing two dice

D. picking a red card from a standard deck of cards; picking a black card from a standard deck of cards

5. The whole numbers from 20 to 40 are each written on separate slips of paper. You draw one slip of paper. Which of the following events would have a probability of zero?

A. picking an even number less than 30
B. picking an odd multiple of 15
C. picking an odd factor of 75
D. picking a prime number

6. A coin-toss game at the carnival uses the board design shown, which is made of two squares. To win, players must land a coin in the shaded area. What is the probability of winning the game, expressed to the nearest percent?

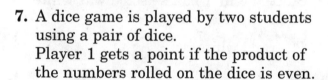

7. A dice game is played by two students using a pair of dice.
Player 1 gets a point if the product of the numbers rolled on the dice is even.

Player 2 gets a point if the product of the numbers rolled on the dice is odd. The player with more points after 20 rounds wins.
Is the game, as outlined, fair or not? Explain.

8. Santos rolled a die 600 times. The results are shown below.

Odd	Even
252	348

a. Calculate the experimental probability for rolling odd as shown by the result given.
b. Determine the theoretical probability for rolling an odd number on a die.
c. Compare the experimental and theoretical probabilities.
d. What could Santos have done to see if the experimental results would come closer to the theoretical results?

---

3 A 2 Probability of Compound Events

Compound events consist of two or more events—for example, rolling a 5 on a die AND picking a red card from a standard deck. If the outcome of one event does not affect the outcome of the other event, the events are **independent**. For two independent events,

$$P(A \text{ and } B) = P(A) \times P(B).$$

If the outcome of one event affects the outcome of the other event, the events are **dependent**. For two dependent events,

$$P(A \text{ and } B) = P(A) \times P(B \text{ after } A \text{ occurs}).$$

Examples
Independent events: The probability of rolling a 5 on a die and picking a red card from a standard deck is

$$P(5 \text{ and red}) = P(5) \times P(\text{red}) = \frac{1}{6} \times \frac{1}{2} = \frac{1}{12}.$$

Dependent events: The probability of picking two letters from the alphabet and having them both be vowels (including Y as a vowel) is

$$P(\text{vowel and vowel}) = \frac{6}{26} \times \frac{5}{25} = \frac{30}{650} = \frac{3}{65}.$$

 # Model Problem

1. Suppose a number cube is rolled twice. What is the probability that an odd number will occur both times?

Solution   Since the first and second rolls of the number cube are independent of each other,
$P(\text{rolling 2 odd numbers}) =$
$P(\text{first roll odd}) \times P(\text{second roll odd}).$

outcomes: 1, 2, 3, 4, 5, 6   total 6

odd number: 1, 3, 5   total 3

$P(\text{odd}) = \frac{3}{6} = \frac{1}{2}$

$P(\text{rolling 2 odd numbers}) = \frac{1}{2} \times \frac{1}{2} = \frac{1}{4}$

2. A jar contains 3 red balls, 2 white balls, and 1 green ball. What is the probability of picking two white balls if the first ball is not replaced?

Solution

$P(\text{first white}) = \frac{2}{6} = \frac{1}{3}$

Since the first ball selected was white and not replaced,

$P(\text{second white}) = \frac{1}{5}$

$P(\text{two whites}) = \frac{1}{3} \times \frac{1}{5} = \frac{1}{15}$

Probability of *A* or *B*

Sometimes you want to find the probability that either of two events will occur. This calculation depends on whether or not the events are *mutually exclusive*. Mutually exclusive events cannot occur at the same time. An example of mutually exclusive events would be rolling a die and getting a 4 or getting an odd number.

For mutually exclusive events,

$$P(A \text{ or } B) = P(A) + P(B).$$

An example of *non-mutually exclusive* events would be rolling a die and getting a 3 or getting an odd number. Notice that both can occur at the same time. To calculate the probability of non-mutually exclusive events, you must subtract $P(A \text{ and } B)$ from the sum of $P(A) + P(B)$ to account for the overlapping of the sample space.

For non-mutually exclusive events,

$$P(A \text{ or } B) = P(A) + P(B) - P(A \text{ and } B).$$

For the spinner shown, find the following probabilities:

a. $P(2 \text{ or } 5)$
b. $P(\text{multiple of 2 or multiple of 3})$

Solution

a. Since obtaining a 2 and obtaining a 5 are mutually exclusive,

$$P(2 \text{ or } 5) = P(2) + P(5) = \frac{1}{8} + \frac{1}{8} = \frac{2}{8} = \frac{1}{4}.$$

b. If the spinner lands on 6, you have a multiple of both 2 and 3. The events are non-mutually exclusive. Observe that the sample spaces overlap:

multiples of 2: {2, 4, 6, 8}
multiples of 3: {3, 6}

$P(\text{mult. of 2 or mult. of 3}) = P(\text{mult. of 2}) + P(\text{mult. of 3}) - P(\text{mult. of 2 and 3})$
$$= \frac{4}{8} + \frac{2}{8} - \frac{1}{8} = \frac{5}{8}$$

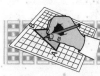 **PRACTICE**

1. A coin is tossed and a die with numbers 1–6 is rolled. What is $P(\text{heads and 3})$?

   A. $\frac{1}{12}$ B. $\frac{1}{4}$ C. $\frac{1}{3}$ D. $\frac{2}{3}$

2. Two cards are selected from a deck of cards numbered 1 through 10. Once a card is selected, it is not replaced. What is $P(\text{two even numbers})$?

   A. $\frac{1}{4}$ B. $\frac{2}{9}$ C. $\frac{1}{2}$ D. 1

3. Which of the following is NOT an example of independent events?

   A. rolling a die and spinning a spinner
   B. tossing a coin two times
   C. picking two cards from a deck with replacement of first card
   D. selecting two marbles one at a time without replacement

4. A club has 25 members, 20 boys and 5 girls. Two members are selected at random to serve as president and vice president. What is the probability that both will be girls?

   A. $\frac{1}{5}$ B. $\frac{1}{25}$ C. $\frac{1}{30}$ D. $\frac{1}{4}$

5. A jar contains two white marbles and one black marble. One marble is drawn randomly and then replaced. A second marble is drawn. What is the probability of drawing a white and then a black?

A. $\frac{1}{3}$ B. $\frac{2}{9}$ C. $\frac{3}{8}$ D. $\frac{1}{6}$

6. Morgan rolls a pair of dice. What is the probability that she obtains a sum that is either a multiple of 3 or a multiple of 4?

A. $\frac{5}{9}$ B. $\frac{7}{12}$ C. $\frac{1}{36}$ D. $\frac{7}{36}$

7. Greg rolls a pair of dice. What is the probability that he obtains a sum of 2 OR 12?

8. Jerry never pairs his socks after doing laundry. He just throws the socks into the drawer randomly. If the drawer contains 14 white socks and 12 gray socks, what is the probability he will select a pair of gray socks when selecting two socks randomly? Give your answer as a decimal rounded to the nearest thousandth.

9. Find the probability of spinning red AND even given the spinners pictured.

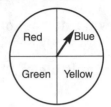 

10. The first spinner shown has four congruent sections. The second spinner has three congruent sections. If you spin each spinner once, what is the probability of obtaining a product of 80?

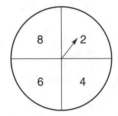 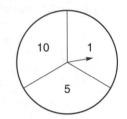

11. Events A and B are independent. The probability of event A occurring is $\frac{3}{5}$ and the probability of event B not occurring is $\frac{2}{3}$. What is $P(A \text{ and } B)$?

12. Suppose E and F are independent events. The probability that event E will occur is .7 and the probability that event F will occur is .6.

   a. Find the probability of E and F both occurring.
   b. Explain why the answer should be less than each of the individual probabilities.

Assessment Macro A

1. One marble is randomly drawn from a jar containing two white marbles and one black marble, and then NOT REPLACED. A second marble is drawn. What is the probability of drawing a white one and then a black one?

   A. $\frac{1}{3}$ B. $\frac{2}{9}$

   C. $\frac{3}{8}$ D. $\frac{1}{6}$

2. The Web Explorers club has 21 members, 12 girls and 9 boys. They decide to pick a president and a vice president randomly. What is the probability that two girls will be selected?

   A. $\frac{1}{4}$ B. $\frac{9}{35}$

   C. $\frac{11}{35}$ D. $\frac{4}{7}$

3. In rolling two dice, the probability of obtaining a sum of 12 (the largest possible sum) is $\frac{1}{36}$. What is the probability of obtaining the largest possible sum when you roll five dice?

   A. $\frac{5}{36}$ B. $\frac{1}{1,296}$

   C. $\frac{1}{216}$ D. $\frac{1}{7,776}$

4. Marla and Darla are twin sisters. Grandma calls them on the telephone once a week on a random day. Whenever the telephone rings, there is an equal chance that Marla or Darla will answer. What is the probability that Grandma will call on a Wednesday and that Darla will answer the telephone?

   A. $\frac{1}{49}$ B. $\frac{1}{14}$

   C. $\frac{1}{7}$ D. $\frac{1}{2}$

5. The Robinson family plans to have three children. Assume that the probability of having a boy and the probability of having a girl are the same. What is the probability that the oldest child will be a girl?

   A. $\frac{1}{3}$ B. $\frac{1}{8}$

   C. $\frac{1}{2}$ D. $\frac{3}{8}$

6. Which of the following pairs of probabilities would NOT represent complementary events?

   A. $\frac{1}{5}$ and $\frac{4}{5}$ B. $\frac{1}{2}$ and $\frac{1}{2}$

   C. .01 and .9   D. .01 and .99

7. Dimitri is asked to find the probability that a die will show an even number or a number greater than 1 on a single roll of the die. He adds the individual probabilities and comes up with a probability of $\frac{8}{6}$. Explain why this result is NOT reasonable.

8. The digits 2, 3, 4 are used to form three-digit whole numbers, without any repetition of digits. What is the probability that a randomly selected number, from among all those generated by the above, is NOT an even number?

9. For the spinner shown, the probability of landing on each color is 25% or $\frac{1}{4}$.

Draw a spinner so that the probabilities (P) would be as follows:

P(red) = .5
P(yellow) = .25
P(orange) = .2
P(brown) = .05

10. A box contains 15 marbles: 6 green, 4 red, and 5 blue. Name a compound event involving picking marbles from the box that would have a probability of $\frac{4}{15} \times \frac{3}{14} \times \frac{2}{13}$.

11. Three dice are rolled. Obtaining a sum of 3 is not very likely. What other sum would have the same probability as a sum of 3?

12. The Sunware Corporation surveyed its 400 employees before changing the company dress code. The results of the survey are shown in the table.

Stronger Dress Code			
	For	**Against**	**Total**
Men	80	120	200
Women	110	90	200
Total	190	210	400

Find the probability that a person chosen at random from among the 400 employees is a man who is against a stronger dress code.

Open-Ended Questions

13. a. Find the probability that an object landing randomly on the figure will land in the shaded circle. Express the probability as a percent.

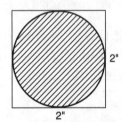

b. Would the probability change if we used this double diagram? Explain your answer.

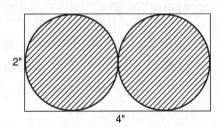

14. In a family of three children there are 8 possibilities for the boy-girl combinations, such as BBG, GBG, and so on.

a. The probability of all boys equals the probability of all girls. What would each of these be?

b. The probability of 2 boys and 1 girl is equal to $\frac{3}{8}$. Explain why it is $\frac{3}{8}$ and not $\frac{1}{8}$.

c. For this family of three children, what other situation would have a probability equal to $\frac{3}{8}$?

d. If the family had four children, how would the answer to part a change?

MACRO B

Collect, organize, represent, analyze, and evaluate data.

3 B 1 Statistical Measures

A set of data can be described by using the *mean*, *median*, *mode*, or *range*. The mean, the median, and the mode are called **measures of central tendency** because they describe the center of a set of data.

- The **mean** is the arithmetic **average**. It is found by dividing the sum of the values by the number of values.
- The **median** is the middle value when the values are listed in order. (*Note:* If the set contains an even number of values, the median is the average of the two values in the middle.)
- The **mode** is the value occurring most frequently.
- The **range** is the difference between the largest value and the smallest value.
- The **frequency** of a value is the count of the value—that is, how many times the value appears.

Example The data set contains the ages of reporters for the city newspaper.

$$36 \qquad 36 \qquad 36 \qquad 43 \qquad 43 \qquad 51 \qquad 54 \qquad 58 \qquad 61$$

mean = 418 ÷ 9 = 46.4   mode = 36
median = 43   range = 61 − 36 = 25

The frequency of 36 is 3, of 43 is 2, and of the other values is 1.

 Model Problem

---

1. Debbie has the following scores on five math tests:

88	84	80	90	84

What score must she get on the sixth test in order for her average to fall between 85 and 87?

Solution The sum of the first 5 grades is 426.

If the mean of the 6 grades is to be between 85 and 87, then the sum of the 6 grades would be between
$$6(85) = 510 \quad \text{and} \quad 6(87) = 522.$$

To find out how many points Debbie needs, find the difference between what she has after 5 tests and what she needs to have after the 6th test.

$$510 - 426 = 84$$
$$522 - 426 = 96$$

Answer To bring the mean between 85 and 87, Debbie must score between 84 and 96 on the 6th test.

---

2. Give a set of five scores such that the data would have

   a. a median of 60
   b. a mode of 52
   c. a mean of 65

Solution

a. __ __ 60 __ __

If the median for five scores is 60, the third score must be 60.

b. 52 52 60 __ __

Since the mode is 52, the lowest two scores must each be 52.

c. Since the mean is 65, the sum of the five scores must be

$$5 \times 65 = 325.$$

With 164 as the sum of the first three scores, the highest two scores must total

$$325 - 164 = 161.$$

Hence, the remaining two scores must be any two distinct numbers above 60 that total 161. For example, one solution is: 52, 52, 60, 80, 81. There is more than one correct answer.

PRACTICE

1. During a baseball season, the National League home run champion had the following home run statistics by month:

April	May	June	July	Aug.	Sept.	Oct.
5	13	7	11	6	8	6

Which month contains the median for the player's home run statistics?

A. June   B. July
C. August   D. September

2. For each number shown in the box, the units digit is hidden. Which of the following could NOT be the mean of the set?

8△	8△	7△	9△	8△

A. 85   B. 82   C. 80   D. 71

3. The following data represent morning temperatures for the month of July in Washington, D.C. What are the mean and median of the data?

July Temperatures (°F)						
Sun.	Mon.	Tues.	Wed.	Thurs.	Fri.	Sat.
				89	87	85
88	93	93	93	89	90	91
90	89	88	89	87	90	91
92	92	92	90	89	87	85
86	84	84	83	85	86	88

A. mean 84.3, median 91
B. mean 82.9, median 90
C. mean 88.5, median 89
D. mean 83.9, median 88

4. For the given scores, the mean is 40.

Scores: 20, 30, 40, 50, 60

If the 20 is changed to a 17, which of the following changes would keep the mean at 40?

A. Change the 50 to a 47.
B. Change the 60 to a 57.
C. Change the 50 to a 53.
D. Change the 30 to a 27.

5. For a set of 6 scores, the following can be noted:

> Score #1 is 6 points below the mean.
> Score #2 is 10 points below the mean.
> Score #3 is 4 points below the mean.
> Score #4 is equal to the mean.

Which of the following could be TRUE about the remaining two scores?

A. Scores #5 and #6 are both equal to the mean.
B. Score #5 is 12 points above the mean and Score #6 is 8 points above the mean.
C. Score #5 is 10 points above the mean and Score #6 is 10 points below the mean.
D. Score #5 is 20 points above the mean and Score #6 is 4 points above the mean.

6. Which of the following statements will always be TRUE?

  I. The mode is always close to the median.
  II. The median is sometimes not included in the data.
  III. The mean is always included in the data.

A. I only   B. II only
C. III only   D. I and II only

7. The list shows the ticket prices for different seats at a concert:

$40 $45 $50 $58 $60 $67 $80 $90

If an additional ticket price of $16 is added to the list, which measure of central tendency is affected most?

8. Eduardo had the following test scores in his science class:

  90   73   86   89   97

What score must he get on the sixth test in order for his average to turn out to be 89?

9. The mean for a set of 5 scores is 60. The mean for a different set of 10 scores is 90. What is the mean for all 15 scores?

10. Give three different values for x so that 80 would be the median.

Score	Number of Students
90	4
85	2
80	3
75	x
70	4

11. Mr. Abbott asked his students to use the following data to find average test scores.

Mr. Abbott's Classes		
Period	Number of Students	Test Average
1 Hygiene	20	80
2 Hygiene	20	70
6 Health Science	30	84
7 Health Science	10	80

In computing the average test score for the Hygiene classes, William suggested that Mr. Abbott take the average of 80 and 70 to get 75. For the Health Science classes, however, using the same approach gives a wrong result of 82. Explain why the first average (75) was correct, but the second average (82) was NOT correct. Find the correct average for the Health Science classes. Explain your approach.

3 B 2 Sampling

The entire group of objects or people involved in a statistical study is called a **population**. However, it is usually impossible to study every member of a large group. For this reason, statistical studies usually look at a smaller group, called a **sample**, which represents the population. **Sampling** is the process of choosing the sample population. It is important to make sure the sample collected is unbiased and represents the entire population. A random sample will give everyone or everything the same chance of being selected.

 Model Problem

The Cake Box surveyed people about the type of frosting they preferred on cakes. Use the results to predict how many of the 952 students at Washington Middle School would choose whipped cream.

Type of Frosting	Percent
Whipped cream	55%
Butter cream	40%
No preference	5%

Solution Since 55% of the sample preferred whipped cream, finding 55% of 952 students allows you to make the prediction.

Answer 0.55(952) = 523.6 or approximately 524 students

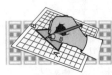 **PRACTICE**

1. Which of the following samples is an example of an unbiased survey?

   A. a random sample of 500 teens in the Northeast to determine the favorite music group for teens ages 13–15

   B. a random sample of 500 men over 50 years of age to determine which brand of vitamins men over 50 prefer

   C. a random sample of 250 women aged 18–35 to determine the favorite brand of ice cream of people 18–35

   D. a random sample of 150 zoo visitors to determine if taxpayers feel that federal money should be used to help run the zoo

2. A sporting goods company surveyed 800 baseball players to see what type of bat they preferred. Aluminum bats were preferred over wood by 300 players. Which statement is true?

   A. More than $\frac{1}{2}$ of the players surveyed preferred aluminum.

   B. More than 40% of the players surveyed preferred aluminum.

C. More than 75% of the players surveyed do not prefer aluminum.

D. More than $\frac{1}{3}$ of the players surveyed prefer aluminum.

3. An office supply store surveyed a group of 200 students to determine their preference for backpack colors. Backpacks come in green, black, blue, and red. Based on the survey results, the store will determine the color distribution for its order of 1,000 backpacks. If 75 students chose black, 25 chose red, 40 chose green, and the rest chose blue, how many green backpacks will the store order?

4. A television rating service found that 945 households out of a sample of 3,340 households watched the Super Bowl. Estimate to the nearest million how many of the 94 million households with a television watched the Super Bowl.

5. Zoologists captured 400 wildebeests, tagged them, and released them back into the same region. Later that season, the wildebeest population was sampled to estimate the size of the population that lived in the region. If a sample of 150 wildebeests contained 45 tagged wildebeests, what would a good estimate be for the wildebeest population in the region?

3 B 3 Data Displays

Data can be organized and displayed by using a variety of different graphs. Tables, charts, matrices, and spreadsheets are also commonly used to display data. The type of graph or device used is determined by the nature of the data and what the data are intended to communicate.

A **circle graph** is used to compare parts of a whole. It is sometimes called a **pie chart**.

Profits From Local Carnival

A **bar graph** compares amounts of quantities.

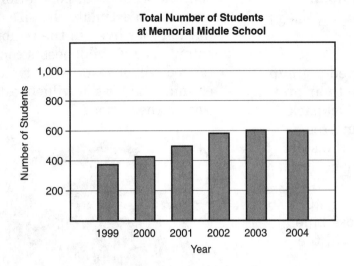

A **pictograph** also compares amounts. A symbol is used to represent a stated amount.

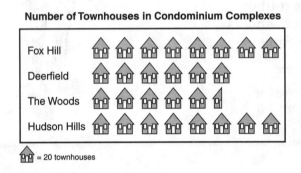

A **histogram** is a bar graph used to show frequencies. In a histogram, the bars, which usually represent grouped intervals, are adjacent.

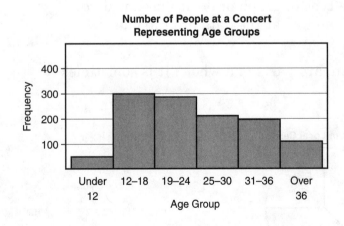

A **line graph** shows continuous change and trends over time.

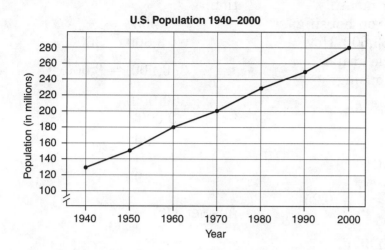

A **line plot** is another way to organize frequency data. A line plot is a picture of the data on a number line corresponding to the range of the data.

Ages of the Haversham Cousins

As is shown in the example above, in a line plot you place an X (or a dot) above the appropriate number to indicate each occurrence.

Model Problem

1. You must construct a graph showing the trend in the price of a gallon of gasoline each month over a five-year period. Which of the following types of displays would be most appropriate?

   A. circle graph
   B. histogram
   C. line plot
   D. line graph

Solution Since the situation depicts changing prices over a period of time, the most appropriate display would be the line graph.

A circle graph is not appropriate since you are not looking at parts of a whole.

A histogram is not appropriate since you are not comparing intervals of prices.

A line plot is not appropriate since you are not organizing frequency data.

2. The Galvins want to make a circle graph representing their monthly budget. They spend $900 on housing out of a total budget of $3,600. How many degrees of the circle graph should represent their housing expenses?

Solution

Method 1   Find the percent.

$$\frac{\text{housing expenses}}{\text{total buget}} = \frac{900}{3,600} = \frac{1}{4} = 0.25$$

Therefore, they need 25% of 360°, which is 90°.

Method 2   Set up and solve a proportion.

$$\frac{900}{3,600} = \frac{x}{360}$$

$$360(900) = 3,600x$$

$$\frac{360(900)}{3,600} = x$$

$$\frac{900}{10} = x$$

$$90 = x$$

Answer   90°

Spreadsheets

A **spreadsheet** also organizes data in rows and columns. Typically, spreadsheets are accessed through computers. In working with a spreadsheet, one considers the spreadsheet to be a large rectangular array of boxes or cells, each of which is identified by a unique address. The address consists of a letter to indicate the column in which the cell is located and a number to indicate the row in which the cell is located. A spreadsheet giving dimensions (in inches) for different rectangular solids might look like:

	A	**B**	**C**	**D**
1	**Length**	**Width**	**Height**	**Volume**
2	5	3	10	150
3	20	10	10	2,000
4	10	10	10	1,000

The address A3 refers to the cell located in column A, row 3. In the spreadsheet shown the number 20 is at that address.

The power of the spreadsheet lies in the fact that each cell can contain a numerical value determined either by direct entry of the value from the keyboard or by a mathematical formula using information obtained from cells anywhere in the spreadsheet under consideration.

For example, the value in D2 was calculated using the formula A2 × B2 × C2. If the values in any of the cells A2, B2, or C2 were changed, then the value in D2 would also change.

Model Problem

The population of two towns remained stable for many years. During that period, Oak Brook's population was approximately 25,000 and Westville's population was approximately 40,000. Suddenly, 20% of Oak Brook's population started moving to Westville each year, while the rest remained in Oak Brook. At the same time, 15% of Westville's population started moving to Oak Brook each year, while the rest remained in Westville. Refer to the spreadsheet for population figures (rounded to the nearest whole number).

a. What is the change in Oak Brook's population over five years?
b. What is the percent of increase in population for Oak Brook over that period?
c. What formula was used to generate the values in A4, B4, and C4?

	A	B	C
		Oak Brook	Westville
1			
2	Year Number	Population	Population
3	0	25,000	40,000
4	1	26,000	39,000
5	2	26,650	38,350
6	3	27,073	37,928
7	4	27,348	37,653
8	5	27,526	37,475

Solution

a. To determine the change in Oak Brook's population, find the difference between the values in B3 and B8.
$$27{,}526 - 25{,}000 = 2{,}526$$

b. The percent of increase = increase ÷ original population.
$$2{,}526 \div 25{,}000 = 10\% \text{ approximately}$$

c. The formulas used were:

For A4, add one to the prior year: $A4 = A3 + 1$.

For B4, 80% of Oak Brook's population stayed and 15% of Westville's is added, so
$$B4 = 0.80 \times B3 + 0.15 \times C3.$$

For C4, 85% of Westville's population stayed and 20% of Oak Brook's is added, so
$$C4 = 0.85 \times C3 + 0.20 \times B3.$$

PRACTICE

1. Which of the following types of graphs would NOT be an appropriate representation to depict the way a family budgets its September income?

   A. bar graph   B. pictograph
   C. circle graph   D. line graph

2. Using the given pictograph, determine what percent of the total number of cars sold at Thrifty's in September Dan sold.

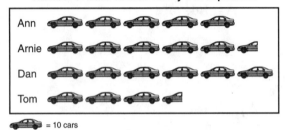

Number of Cars Sold at Thrifty's in September

Ann
Arnie
Dan
Tom

= 10 cars

   A. 60%   B. 40%   C. 30%   D. 20%

3. While he is training for a triathlon, Tim spends $2\frac{1}{2}$ hours exercising each day. Use the given circle graph to determine what percent of Tim's exercise program is devoted to running.

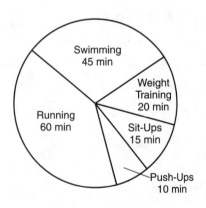

Swimming
45 min

Weight
Training
20 min

Running
60 min

Sit-Ups
15 min

Push-Ups
10 min

   A. 25%   B. 40%   C. 60%   D. $66\frac{2}{3}$%

4. For which of the following situations would it NOT be appropriate to use a line graph to represent the data?

A. Show the population of the U.S. from 1900 to 1990.
B. Show the sale of CD's during a five-year period.
C. Show survey results of how students spend one hour of their time.
D. Show the heating time for water at various altitudes.

5. If the data in the table were represented in a circle graph, what would be the degree measure of the section representing pizza?

Favorite Lunch Survey	
Main Course	**Number of Students**
Chicken Fingers	40
Hamburgers	44
Hot Dogs	45
Pasta	21
Pizza	50

   A. 25   B. 50   C. 75   D. 90

6. The spreadsheet represents data collected from five students who were asked five questions about changes proposed to their school. A score of 1 indicated a least favorable response, while a score of 5 indicated a most favorable response. To find out the average response for question 4, which cell would you look in?

	A	B	C	D	E	F
1	Name	Q #1	Q #2	Q #3	Q #4	Q #5
2	Alice	2	1	3	4	5
3	Bill	3	5	2	1	4
4	Dave	4	5	3	2	1
5	Jennifer	2	3	4	1	5
6	Franco	5	2	3	1	4
7						
8	Average	3.2	3.2	3.1	1.8	3.8

   A. E8   B. B8   C. C8   D. D8

7. If the circle graph represents the city of Metropolis, which has 42,000 homes, how many homes are NOT heated by natural gas?

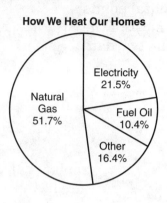

How We Heat Our Homes

Electricity 21.5%

Natural Gas 51.7%

Fuel Oil 10.4%

Other 16.4%

8. The South Side Middle School's annual fundraiser involved the sale of tins of cookies. The fifth grade sold 2,586 tins of cookies; the sixth grade sold 3,014 tins; the seventh grade sold 3,274 tins; and the eighth grade sold 3,326 tins. Construct a graph to show how the grades compared in amounts of cookies sold. Explain why you selected that type of graph to represent the data.

Open-Ended Questions

9. Test scores in a biology class are as follows:

83, 78, 94, 93, 87, 86, 83, 94, 99, 90, 87, 79, 65, 87, 93, 96, 88, 84, 82, 93, 85.

a. Construct a line plot for the data.
b. State the median score for the data.

10. This graph shows the percentages of pickle buyers who selected various types of pickles.

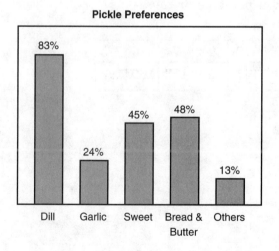

Pickle Preferences

83%

24%

45%

48%

13%

Dill   Garlic   Sweet   Bread & Butter   Others

a. Explain why the data cannot be used to construct a circle graph.
b. Explain what is wrong or misleading in the given graph.

11. The line plot displays scores on an 80-point mathematics test.

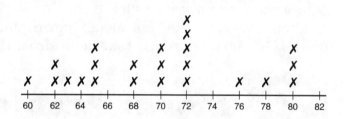

a. Which measure of central tendency (mean, median, or mode) is most easily observed from this line plot? Explain why.
b. What is the median test score? Explain how you find it from this line plot.
c. Suppose the teacher gives a make-up test and must add the following six scores to the line plot:
72, 76, 76, 78, 80, 80.
(i) Does the median change? If so, what is the new median?
(ii) Does the mode change? If so, what is the new mode?
(iii) In a general way, how does the mean change and why? (Do not actually calculate the mean.)

3 B 4 Relationships Involving Data

A **scatter plot** is a graph used to show a relationship or **correlation** between sets of data. In a scatter plot, we plot the data as ordered pairs, represented by unconnected points. The pattern of the data points shows the correlation, if any, between the two data sets. If most of the data points are clustered together along an imaginary line, the two data sets are correlated.

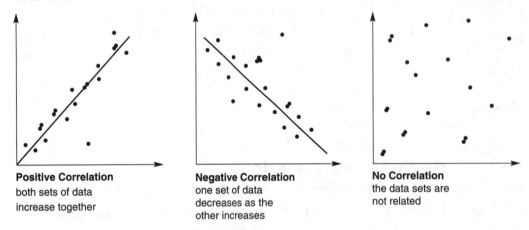

Positive Correlation
both sets of data
increase together

Negative Correlation
one set of data
decreases as the
other increases

No Correlation
the data sets are
not related

A **line of best fit** or **trend line** can be drawn near where most of the points cluster on a scatter plot. If the line slopes upward, a positive correlation exists. If the line slopes downward, a negative correlation exists. **Outliers** are points that lie far from the overall linear pattern.

Model Problem

1. Plot the data given in the table on a graph. Draw the trend line that best fits the data. Does the graph show a positive or negative correlation?

Height (in.)	60	62	63	65	68	69	70	70	72	74	75	75
Weight (lb)	120	122	125	130	132	142	158	147	150	152	160	156

Solution

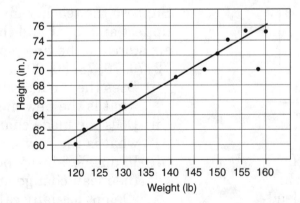

Answer   The trend line slopes upward, showing a positive correlation.

2. Describe the correlation that would exist in each of the following:

   a. outside temperature vs. the amount of heating oil used by a furnace

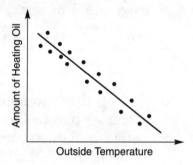

   b. hours worked vs. earnings

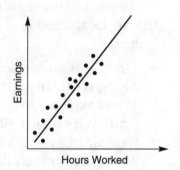

   c. shoe size vs. score on the math test

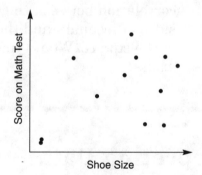

Solution

a. The scatter plot suggests a line that slopes downward to the right. As the outside temperature increases, the amount of heating oil needed decreases. This is a negative correlation.
b. As the hours worked increase, the earnings increase. Therefore, a positive correlation exists.
c. The points are spread out. There appears to be no correlation between shoe size and math scores.

1. For which of the following situations would you expect the scatter plot to show a negative correlation?

    A. the number of students in a high school and the average temperature of the city
    B. the age of a car and the resale value
    C. the price of an item and the amount of tax on the item
    D. the speed of a car and the distance traveled in a fixed time

2. Which scatter plot shows a positive correlation?

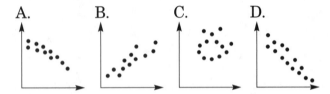

A.   B.   C.   D.

3. For which of the following situations would you expect the scatter plot to show no correlation?

    A. miles driven and gallons of gas used
    B. driving speed and driving time on a 5-mile stretch of highway
    C. number of pages in a book and number of copies sold
    D. oven temperature and cooking time for a 12-pound turkey

4. Display the data shown in the table in a scatter plot and describe the type of correlation present.

x	10	10.5	11	12	9.5	13	10.5
y	47	35	63	22	55	27	9

5. Two students in a geometry class made a scatter plot to show the relationship between diameter and circumference of circular objects. For each object (such as the top of a coffee can), they plotted the points (diameter on the horizontal axis and circumference on the vertical axis) and drew the line of best fit.

    a. Draw a possible scatter plot to go along with this situation.
    b. What type of correlation did they find? Why does this make sense?
    c. Explain why you would not expect your data to include any outliers.

6. Explain why you would NOT need to gather data and draw a scatter plot in order to determine the type of correlation between the length of the sides of a square and the perimeter of the square. What type of correlation exists?

3 B 5 Evaluating and Interpreting Data

To use a graph to interpret data:

- Pay attention to the scale. Check to see if the scale has a broken line between zero and the first interval.
- Know what the numbers mean and be aware of units.
- Read the title and the labels on the axes.
- Check graphs with multiple lines or bars for relationships between points.
- Be able to make predictions about the relation that *go beyond* what is displayed.

To use a table to interpret data:

- Read the title and labels.
- Know what the numbers mean and be aware of units.
- Be able to estimate values *between* given entries.

When interpreting data, be alert for misuses and abuses of statistics. Statistics can be misleading in the following circumstances:

- The scale used is inappropriate to display data.

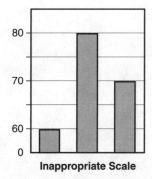
Inappropriate Scale

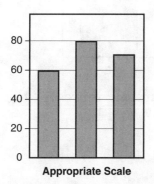
Appropriate Scale

- The measure of central tendency used is not a good description of the data. In a small company of six people, the salaries of individuals are $80,000, $20,000, $25,000, $19,000, $22,500, and $23,500. The mean salary is about $32,000. Using the mean would be inappropriate to describe the set, since four out of five people earn less than that. A better descriptor of the data would be the median, $23,000.

- The titles or labels on the axes or on a table are insufficient.

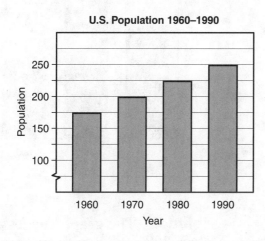

The vertical axis does not indicate that the numbers are in millions.

- The visual display of the data creates an impression different from what the data suggests.

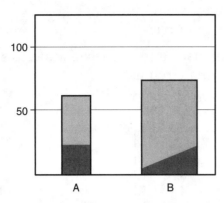

The visual display does not accurately represent the numerical relationship between A and B.

 # Model Problem

1. The graph displays the average monthly temperatures for two different years. How do the temperatures for Year A compare to those for Year B? Explain.

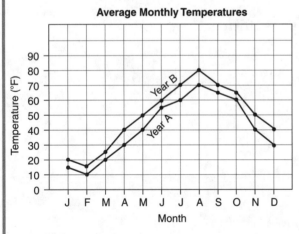

Average Monthly Temperatures

Solution The graph of the average monthly temperatures for Year B is above the graph for Year A, indicating that the average monthly temperatures for Year B were greater than those for Year A. Since the two graphs never intersect, it is clear that at no time did a month in Year B record a temperature less than or equal to any in the same month for Year A.

2. What impression does the graph give? How is this impression created?

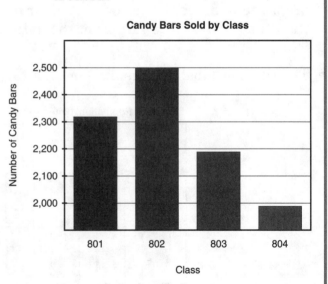

Candy Bars Sold by Class

Solution The graph suggests that class 802 sold significantly more candy bars than class 804. The scale gives the same amount of space to the interval from 0 to 2,000 as to the intervals of one hundred. This is inappropriate.

PRACTICE

1. The following graph shows the average monthly 6:00 A.M. and 6:00 P.M. temperatures (°F) recorded at Newark Airport.

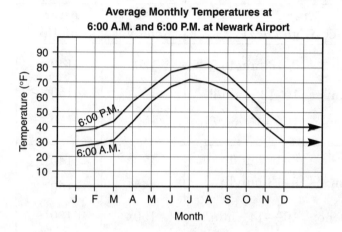

a. How do you know that the graph does NOT show a situation in which the September temperatures at 6:00 A.M. and 6:00 P.M. were the same?

b. According to the graph, was there any month when the average 6:00 A.M. temperature was greater than the average 6:00 P.M. temperature? Explain your response.

2. The circle graphs show how Sally and Michele spend their earnings. How is it possible that Michele can spend a greater dollar amount on recreation than the dollar amount spent by Sally?

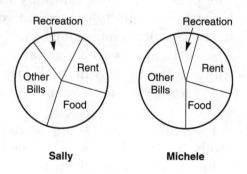

3. The graph shows the volume of sales of cassette tapes and compact discs (CD's) at a local music store over a six-year period. Explain the trend shown separately in Line A and Line B. Discuss the significance of the point of intersection of Line A and Line B.

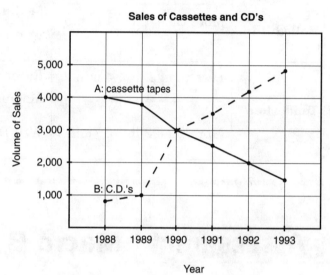

4. The graph is intended to compare morning coffee sales at two convenience stores in the same town. Is the visual message depicted in the graph accurate? Explain your response.

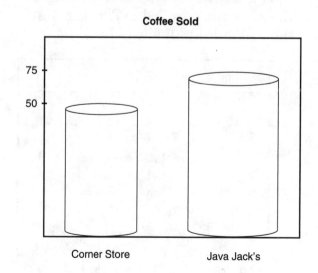

5. Mrs. Mendez is giving a dinner party. She will serve a $5\frac{3}{4}$ pound standing rib roast that she plans to cook in a microwave oven. She knows that her guests prefer the meat to be cooked medium. Using the information shown on the recipe card, figure out the minimum amount of time (in minutes) needed to cook the roast.

BEEF	Microwave Time				Internal Temperature	
	Step 1 High (100%)		Step 2 Med-High (70%)		At Removal	After Standing
Standing or Rolled Rib	Less than 4 lb	$6\frac{1}{2}$ min	Rare:	9–13 min/lb	120°	140°
	More than 4 lb	$10\frac{1}{2}$ min	Medium:	10–13$\frac{1}{2}$ min/lb	135°	150°
			Well Done:	10$\frac{1}{2}$–15 min/lb	150°	160°
Tenderloin	Less than 2 lb	4 min	Rare:	8–11 min/lb	120°	140°
			Medium:	9–13 min/lb	135°	150°
	More than 2 lb	$6\frac{1}{2}$ min	Well Done:	10$\frac{1}{2}$–14$\frac{1}{2}$ min/lb	150°	160°

Assessment Macro B

1. Consider the set of data: 92, 80, 79, 75, 75, 58, 55. If the 58 and 55 were dropped from the data, which measure would remain unchanged?

   A. median   B. mode
   C. mean   D. range

2. This histogram shows final averages of the students enrolled in Algebra 1 at the North End Middle School.

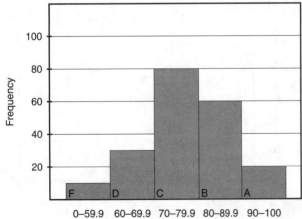

What percent of the students scored B or better in the course?

   A. 20%   B. 40%
   C. 60%   D. 80%

3. For which of the following situations would you use the entire population rather than a sample?

   A. A cook making sauce wants to know if there is enough salt in it.
   B. The president of the company wants to know how many people are going to the company holiday party.
   C. A statistical research company wants to know how many households watched the Academy Awards show.
   D. A medical research group wants to study the blood cholesterol levels of women in the United States over age 50.

4. For which of the following situations would you use the mean to analyze the data?

   A. the preferred color marker students buy

   B. the day of the week on which most students were born

   C. the grades a class received on a test

   D. the difference between the ages of the oldest and youngest students in the class

5. The following graphs depict the student population of Donnybrook Middle School from 1997 to 2004.

I.
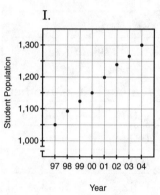

II.
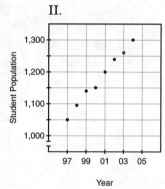

III.
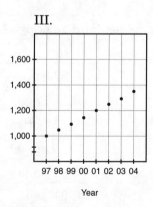

   a. Which graph would you use to convince someone that the student population increased steeply during the period?

   b. Between which two years did the student population show the most growth?

6. Use this bus schedule to answer the question that follows.

Sundays								
HACKENSACK • Sears Main & Anderson Sts.	TEANECK Cedar Lane & Queen Anne Rd.	TEANECK Queen Anne Rd. & DeGraw Ave.	RIDGEFIELD PARK Main & Mt. Vernon Sts.	PALISADES PARK • Morsemere Broad & Columbia Aves.	RIDGEFIELD Traffic Circle	FAIRVIEW Nungessers	NORTH BERGEN Blvd. East & 74th St.	NEW YORK Port Authority Bus Terminal
A.M.	A.M.	A.M.	A.M.	A.M.	A.M.	A.M.	A.M.	A.M.
7.23	7.27	7.32	7.37	7.42	7.44	7.50	7.54	8.10
8.23	8.27	8.32	8.37	8.42	8.44	8.50	8.54	9.10
9.23	9.27	9.32	9.37	9.42	9.44	9.50	9.54	10.10
10.23	10.27	10.32	10.37	10.42	10.44	10.50	10.54	11.10
–	–	–	–	–	–	–	–	P.M.
11.23	11.27	11.32	11.37	11.42	11.44	11.50	11.54	12.10
P.M.	P.M.	P.M.	P.M.	P.M.	P.M.	P.M.	P.M.	–
12.23	12.27	12.32	12.37	12.42	12.44	12.50	12.54	1.10
1.23	1.27	1.32	1.37	1.42	1.44	1.50	1.54	2.10
2.23	2.27	2.32	2.37	2.42	2.44	2.50	2.54	3.10
3.23	3.27	3.32	3.37	3.42	3.44	3.50	3.54	4.10
4.23	4.27	4.32	4.37	4.42	4.44	4.50	4.54	5.10
5.23	5.27	5.32	5.37	5.42	5.44	5.50	5.54	6.10
6.23	6.27	6.32	6.37	6.42	6.44	6.50	6.54	7.10
7.26	7.32	7.37	7.42	7.47	7.49	7.55	7.59	8.15
8.26	8.32	8.37	8.42	8.47	8.49	8.55	8.59	9.15
9.26	9.32	9.37	9.42	9.47	9.49	9.55	9.59	10.15
10.26	10.32	10.37	10.42	10.47	10.49	10.55	10.59	11.15
–	–	–	–	–	–	A.M.	A.M.	A.M.
11.37	11.41	11.46	11.51	11.56	11.58	12.04	12.06	12.24

Mr. and Mrs. Norton are taking a bus from Teaneck to New York City on Sunday in order to see *The Lion King*. The show begins at 3:00 P.M. and it normally takes 20 minutes to go from the bus terminal in NYC to the theater. If they plan to board the bus at the corner of Cedar Lane and Queen Anne Road, what is the latest time they can board the bus?

   A. 12:27 P.M.   B. 1:27 P.M.

   C. 1:32 P.M.   D. 2:27 P.M.

7. Using the spreadsheet shown, what was the approximate percent of increase of $100 after it was invested at the rate of 6% compounded annually for five years?

	A	B	C	D	E
1	Compound Interest				
2	Original Price	Rate	Period/Yr	Year	New Balance
3	$100	6.00	1	1	$106.00
4				2	$112.36
5				3	$119.10
6				4	$126.25
7				5	$133.82
8				6	$141.85
9				7	$150.36

A. 6%   B. 30%   C. 34%   D. 42%

8. The table displays the appraised values of seven houses on the same block.

Appraised Values on Block B	
House Number	Value
326	$350,000
331	$378,500
338	$343,800
343	$355,900
347	$363,500
354	$358,000
365	$786,500

a. What is the difference between the mean and median values of the houses?
b. Which measure, mean or median, is a better representation for the data and why?

9. There are 20 students in a class. Mr. Brock calculated the average grade for the class on a test as 74, but he mistakenly read one student's grade as 50 instead of 90. What will the average grade for that class be when Mr. Brock recalculates using the correct score?

10. A group of biology students captured, tagged, and released a random sample of 50 fish in a lake. Two weeks later, they took a random sample of 27 fish that contained 3 with tags. Estimate the number of fish in the lake.

11. The table below indicates the costs at the Copy Cat copy center. Martina, a member of the center, needs to have 75 copies made of a packet containing 22 pages. She needs the packets collated and stapled. Furthermore, she wants the copies run on three-hole-punch paper. What will be the cost of the job?

Copy Cat Prices		
Number of Sets per Original	Price per Copy Page	
	Member	Non-Member
1	0.05	0.06
2–49	0.04	0.05
50–499	0.03	0.04
500–999	0.025	0.035
1,000+	0.2	0.03
Additional Services (add to copy charges)		
Collating	free	
Stapling	0.02/Staple	
Hand Feeding	0.10/Page	
Reducing	0.25/Setting	
Other Paper Stock (add to copy charges)		
Legal Size Paper	0.01/page	
3-Hole-Punch Paper	0.01/page	
Pastel #20 Paper	0.01/page	
Bright #60 Paper	0.02/page	
Resume Stock	0.05/page	
Card Stock	0.05/page	
Mailing Labels	0.30/page	
Transparencies	0.40/page	
The minimum order is $1.00.		

12. The following double-bar graph compares the number of gold medals won by different countries at the 1988 Winter Olympics in Calgary, Canada, vs. the 1984 Winter Olympics in Sarajevo, Yugoslavia.

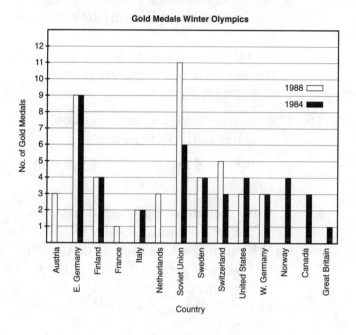

a. Based on the data in the graph, can you assume that since East Germany won the same number of gold medals at each of the two Olympics, it won the same percentage of the gold medals awarded?

b. What country in 1988 won approximately the same percentage of gold medals as East Germany won in 1984?

Open-Ended Questions

13. Miguel's scores in chemistry this quarter are 90, 30, 78, 75, 40, 54, and 70. He decided to use the median grade to report his chemistry grade to his parents. What advantage is there to his using the median rather than the mean?

14. Ms. McNally gave two quizzes on the American Revolution. The average score on both was 85. Scores on the first quiz ranged from 55 to 100, while scores on the second ranged from 73 to 93. Sketch a possible histogram of the scores in each of the two quizzes. Describe how these histograms differ.

15. For each of the given situations, state whether or not you can find the mean from the information given. Explain your response.

a. You know the scores of five individual students.

b. You know the sum of the scores of five individual students.

c. You know the range of the scores of five students.

16. The table lists the life expectancies in years of males as estimated in 1992.

Age in Years	Expected Years Until Death
0	72.2
10	63.1
20	53.4
30	44.2
40	34.9
50	26.1
60	18.2
70	11.8

a. Make a scatter plot of the data on graph paper.
b. Draw a trend line for the scatter plot.
c. Use your trend line to predict the life expectancy of males age 80.

17. Using example sets of five scores, illustrate each of the following situations.

a. If you change one number, the mean of the data will change.
b. If you change one number, the median of the data may or may not change.
c. If you change one number, the mode of the data may or may not change.

18. The annual salaries for five major-league baseball players are:

$120,000 $110,000 $140,000
$120,000 $1,000,000

a. Find the mean for the salaries.
b. What is the median salary?
c. Which of the two measures (mean or median) gives a better indication of the annual salary for the group of baseball players? Explain your response.

Apply the concepts and methods of discrete mathematics to model and explore a variety of practical situations.

3 C 1 Methods of Counting

Counting of discrete or individual items is a major part of the topic of discrete mathematics.

The counting process often involves making an organized list of the items to be counted. For example, to count the number of rectangles of any size in the picture shown at the right, it is necessary to find an organized way of counting each different-size rectangle present.

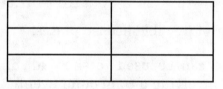

Size a		6
Size b		3
Size c		2
Size d		4
Size e		2
Size f	the entire large rectangle	1
	Total number of rectangles	18

The *counting principle* is an efficient method of finding the total number of ways that a compound event can happen. (A **compound event** consists of two or more events.) The **counting principle** states that if one event can happen in a ways and a second event can happen in b ways, then the two events can occur in $a \times b$ ways. This product can be extended for more than two events.

Example Special license plates require two letters followed by two digits (with all letters and digits possible).

The counting principle says that if there are 26 possible letters for the first letter and 26 for the second letter, and 10 choices for the first digit and 10 choices for the second digit, the product

$$26 \times 26 \times 10 \times 10 = 67{,}600$$

is the total number of possible license plates.

A **tree diagram** can be used to show all of the possibilities in a counting situation involving a compound event.

 # Model Problem

The Lunch Box Cafe offers a soup-and-sandwich lunch. There are two soup choices (tomato or chicken). For the sandwich, there are three bread choices (white, rye, or wheat) and four meat choices (roast beef, ham, turkey, or salami).

How many different lunches are possible?

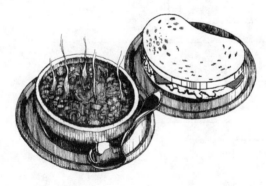

Solution

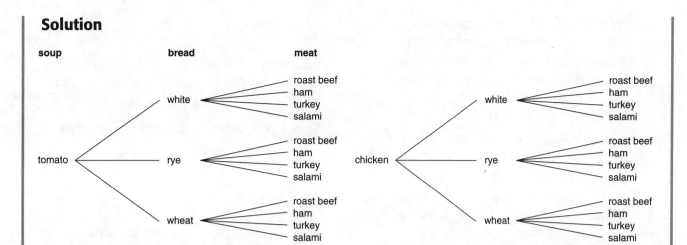

The tree diagram shows 24 final branches. The result (24) is consistent with the use of the counting principle: $2 \times 3 \times 4 = 24$.

Factorial Notation

The counting principle is often applied to problems involving choices and arrangements. For example, an antique car collector wants to display five cars in her driveway. She can choose from 5 cars for the first position, 4 cars for the second position, 3 cars for the third position, and so on. The total number of ways to display the cars is the product of consecutive descending factors, where 1 is the last factor. A notation used for this purpose is called **factorial notation**.

$$5! \text{ (read as 5 factorial)} = 5 \times 4 \times 3 \times 2 \times 1$$

Most scientific and graphing calculators contain a factorial key.
Note: 0! is defined as equal to 1.

 # Model Problem

1. In how many different ways can six students be seated on a six-seat bench?

Solution Any of the 6 students may be seated in the first seat. After a student is seated in the first seat, there are 5 choices for the second seat. For each successive seat, the number of choices decreases by 1. Use the counting principle:

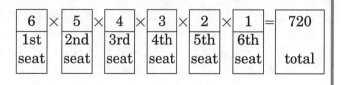

6	×	5	×	4	×	3	×	2	×	1	=	720
1st seat		2nd seat		3rd seat		4th seat		5th seat		6th seat		total

Answer 720 possibilities

2. From the digits 0–9, how many different four-digit numbers are possible if the number must be an odd multiple of 5 with no repeated digits in the number?

Solution Since the number must be an odd multiple of 5, there is only one choice for the units digit: it must be a 5.

For the digit in the thousands place, you cannot use a 5 or a 0, leaving you with 8 possibilities.

For the digit in the hundreds place, you cannot use a 5 or the digit used in the thousands place. As a result, you have 8 possibilities.

It then follows that you will have 7 possibilities for the digit in the tens place.

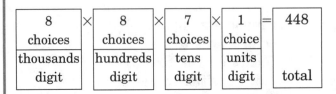

| 8 choices thousands digit | × | 8 choices hundreds digit | × | 7 choices tens digit | × | 1 choice units digit | = | 448 total |

Answer Using the counting principle, you get 448 possible numbers.

3. Find the value of $\frac{8!}{5!}$.

Solution
$$\frac{8!}{5!} = \frac{8 \times 7 \times 6 \times 5 \times 4 \times 3 \times 2 \times 1}{5 \times 4 \times 3 \times 2 \times 1}$$
Therefore, after reducing the fraction,
$$\frac{8!}{5!} = 8 \times 7 \times 6 = 336.$$

Answer 336

4. How many pizzas can be made using 0, 1, 2, 3, or 4 of the following toppings: pepperoni, onion, mushroom, and green pepper?

Solution It is possible to make a systematic list of all the different pizzas, but applying the counting principle by indicating the number of choices regarding each topping may be more efficient.

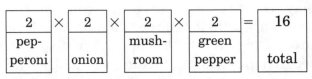

| 2 pepperoni | × | 2 onion | × | 2 mushroom | × | 2 green pepper | = | 16 total |

Using this approach, you are counting 2 possibilities for pepperoni (either it will be on the pie or not), and the same for the other toppings.

Answer There are 16 total possible pizzas.

Permutations

In situations involving counting possibilities, it is necessary to determine if the order of the items matters. A common kind of counting problem is to find the number of arrangements of some or all of a set of objects. Each arrangement is called a **permutation**. The counting principle may be used to calculate the number of permutations. For example, the number of arrangements of four people in line to buy a concert ticket would be:

$$4 \times 3 \times 2 \times 1 = 4! = 24.$$

If only two of the four individuals were arranged in line, the number of permutations would be

$$4 \times 3 = 12.$$

Permutations of n objects at a time:
$$_nP_n = n!$$
Permutations of n objects r at a time:
$$_nP_r = \frac{n!}{(n-r)!}$$

Model Problem

1. Find the value of $_7P_3$.

Solution $_7P_3$ is the number of permutations of 7 things taken 3 at a time.

$$\frac{7!}{(7-3)!} = \frac{7!}{4!} = \frac{7 \times 6 \times 5 \times 4 \times 3 \times 2 \times 1}{4 \times 3 \times 2 \times 1}$$
$$= 7 \times 6 \times 5 = 210$$

2. Using the digits 2, 3, 4, 5, 6, how many 3-digit numbers can be formed if repetition of digits is not permitted?

Solution Since order does matter (245 and 452 are different numbers), this is a straightforward permutation problem.

$$_5P_3 = \frac{5!}{(5-3)!} = \frac{5!}{2!} = \frac{5 \times 4 \times 3 \times 2 \times 1}{2 \times 1}$$
$$= 5 \times 4 \times 3 = 60$$

Note: If it were possible to repeat digits (as in 445 and 333), there would be additional numbers possible. With repetition possible, the solution would be:

$$5 \times 5 \times 5 = 5^3 = 125.$$

Combinations

Combinations involve selecting from among a given number of people or objects where order does not matter. For example, if you were selecting two girls out of a group of four to be on a committee, a listing would show that there are six committees (or combinations) possible:

Alice (A), Bonita (B), Celine (C), Donna (D)

AB	BC	CD
AC	BD	
AD		

The number of combinations is always smaller than the corresponding number of permutations. This is because in a permutation, AB would be counted as different from BA. In a combination, AB and BA would be considered the same.

   As with permutations, there are formulas for calculating the number of combinations.

Combinations of n things r at a time:
$$_nC_r = \frac{n!}{(n-r)!r!}$$

Special cases include:
$_nC_n = 1$ (only 1 combination possible in selecting n things n at a time)
$_nC_1 = n$ (n combinations possible in selecting n things 1 at a time)
$_nC_0 = 1$ (1 combination possible in selecting n things 0 at a time)

 # Model Problem

1. Find the value of $_{10}C_2$.

Solution $_{10}C_2$ is the number of combinations of 10 things 2 at a time.

$\frac{10!}{8! \times 2!} = \frac{10 \times 9}{2}$ (after reducing the fraction because numerator and denominator each contain 8! as a factor)

$= \frac{90}{2} = 45$

2. For a history report, you can choose to research three of the original thirteen American colonies. How many different combinations exist for the colonies you will be researching?

Solution Since you are making a selection where order does not matter, this is a combination problem.

$_{13}C_3 = \frac{13!}{10!3!} = \frac{13 \times 12 \times 11}{3 \times 2 \times 1}$

$= 13 \times 2 \times 11 = 286$ combinations

Answer There are 286 combinations of three colonies.

3. How many pizzas can be made using 0, 1, 2, 3, or 4 of the following toppings: pepperoni, onion, mushroom, and green pepper? (*Note:* This is a repeat of a Model Problem 4 from page 152 with a different solution strategy.)

Solution Assume that a pizza with pepperoni and onion is the same as one with onion and pepperoni; therefore, order does not matter and we can use combinations.

4 toppings $_4C_4 = 1$ (only 1 pizza possible with all toppings)

3 toppings $_4C_3 = \frac{4!}{1!3!} = 4$

2 toppings $_4C_2 = \frac{4!}{2!2!} = 6$

1 topping $_4C_1 = 4$

0 toppings $_4C_0 = 1$

Answer Total number of possible pizzas $= 1 + 4 + 6 + 4 + 1 = 16$.

Pascal's Triangle

Pascal's triangle can be used to solve problems involving combinations. A portion of Pascal's triangle is as follows:

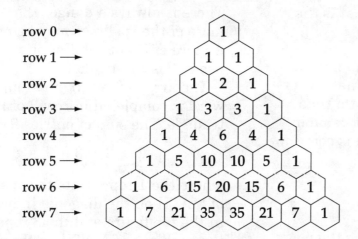

The entries in a given row represent combinations of that number of items. For example, the entries 1, 3, 3, 1 in row 3 represent combinations of three items, such as the letters A, B, and C:

3 items 0 at a time $_3C_0 = 1$ There is only 1 way to pick none of the letters.

3 items 1 at a time $_3C_1 = 3$ There are 3 ways to pick 1 letter: A, B, or C.

3 items 2 at a time $_3C_2 = 3$ There are 3 ways to pick 2 letters: AB, AC, or BC.

3 items 3 at a time $_3C_3 = 1$ There is 1 way to pick 3 letters: ABC.

Additionally, it should be noted that the sum of the entries in a given row involves powers of 2:

$$1 = 2^0$$
$$1 + 1 = 2^1$$
$$1 + 2 + 1 = 4 = 2^2$$
$$1 + 3 + 3 + 1 = 8 = 2^3$$
$$1 + 4 + 6 + 4 + 1 = 16 = 2^4$$

Model Problem

1. A row of Pascal's triangle starts with the numbers 1 and 12. What is the next number in this row?

Solution The 1 and 12 would indicate the 12th row involving combinations of 12 things a certain number at a time. $_{12}C_1 = 12$; therefore, the next value in the row would be $_{12}C_2$, which is equal to $\frac{12!}{10!2!} = 66$.

2. Find the value of $_6C_3$.

Solution In addition to using the formula for combinations, you can simply take the fourth number of the sixth row of Pascal's triangle:

$$1 \quad 6 \quad 15 \quad 20 \quad 15 \quad 6 \quad 1$$

The fourth value is 20.

Note: You use the fourth number because the first number would be $_6C_0$; as a result, the fourth number would be $_6C_3$.

3. Starting at M and proceeding downward on a diagonal either left or right each time, determine how many different paths spell the word MATH.

Solution One approach to finding the solution to this problem is to trace the

spellings of MATH systematically. There is always a danger that one or more of the spellings may be missed. A more efficient solution involves the use of Pascal's triangle.

The array of letters in a four-letter word is completed in row 3 of Pascal's triangle. The sum of entries for row 3 is

$$1 + 3 + 3 + 1 = 8$$

where the 1 represents the spelling of MATH on the diagonal indicated in the top figure and the 3 represents the spellings of MATH on the diagonal paths indicated in the bottom figure. These two figures show the solution for paths to the left of the M at the top of the triangle. Corresponding solutions exist if you move right of the top M. Hence, 1 is added twice and 3 is added twice to obtain the answer of 8.

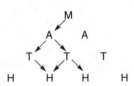

PRACTICE

1. Which of the following involves combinations?

   A. In how many ways can 5 different books be arranged on a shelf?
   B. How many subsets exist from the set {a, b, c, d}?
   C. How many three-digit numbers are possible using the digits 1, 4, 7, and 9?
   D. A school needs to schedule four classes—English, Spanish, math, and science—in the first 4 periods. How many different schedules are possible?

2. Which has the greatest value?

   A. $_8C_6$ B. $_9C_4$
   C. $_9P_4$ D. $_8P_6$

3. April, Brittany, Christina, and Dawn are competing in a race. If there are no ties, in how many different ways can they finish the race?

   A. 6   B. 12   C. 24   D. 36

4. A customer in a computer store can choose one of four monitors, one of two keyboards, and one of four computers. If all of the choices are compatible, how many different systems are possible?

   A. 10   B. 16   C. 32   D. 10!

5. The numbers 1 and 10 are the first two numbers of the tenth row of Pascal's triangle. What is the third number in this row?

6. The track coach needs to select a starter, a second runner, a third runner, and an anchor runner for a four-person relay team. The coach will pick from among Aaron, Byron, Connor, Duncan, Evan, and Finn. How many different four-person relay teams can be made from the six runners?

   A. 24   B. 256
   C. 360   D. 720

7. A school committee consists of the student council president, 5 other students, the principal, and 3 other teachers. In how many ways can a subcommittee be selected if the subcommittee is to consist of the student council president, 2 other students from the committee, the principal, and 1 other teacher from the committee?

8. A restaurant offers a soup-and-sandwich lunch. There are 3 possible soups, 3 breads possible for the sandwich, and k types of meat available for the sandwich. If the tree diagram constructed yields a total of 45 different lunches possible, find the value of k.

9. How many triangles (of any size) are in the diagram?

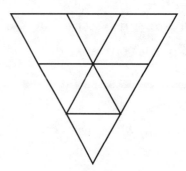

10. Four roads go from town *A* to town *B*. Three roads go from town *B* to town *C*. In addition, there are two roads that go from *A* to *C* without going through *B*. In how many ways can you go from *A* to *C*?

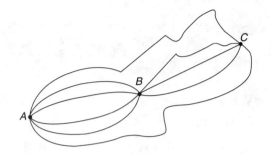

Open-Ended Questions

---

11. The traveling squad for a college basketball team consists of two centers, five forwards, and four guards. The coach is interested in determining the number of ways she can select a starting team of one center, two forwards, and two guards.

  a. Find the number of ways to select one center.
  b. In finding the number of ways to select two forwards, are permutations or combinations used? Explain.
  c. Find the number of ways to select the two starting forwards.
  d. Find the number of ways to select the two starting guards.
  e. Using the answers to parts a, c, and d, would you calculate the number of possible starting teams by finding

  $$a + c + d \quad \text{or} \quad a \times c \times d?$$

  Explain your answer.

12. Suppose a state's license plates have three digits, then a picture of the state bird, and then three more digits.

  a. If any digit (including zero) can go in any position, how many license plates are possible?
  b. If the state expects to use up all of the license plate numbers within the next year, it needs to have a plan to develop more numbers. If it decides to replace the first three digits with letters of the alphabet, how many license plate numbers are now possible? Show your process.
  c. A member of the state transportation board suggested that one additional digit (for a total of 7 digits) would be better than 6 spaces with the three letters and three digits. Is this suggestion accurate? Explain your response.

3 C 2 Networks

The eighteenth-century European town of Königsberg included two islands and seven bridges as shown in the diagram. The question was often asked of whether or not a person could begin anywhere and walk through town crossing all seven bridges without crossing any bridge twice.

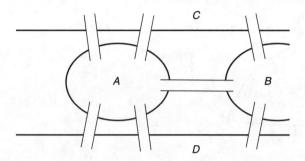

The Swiss mathematician Euler proved that such a walk was impossible. Euler created a geometric model known as a graph or network. In this model, the segments or arcs (edges) represent the bridges and the lettered points (vertices) represent the land regions.

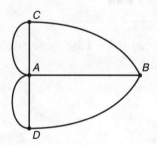

Euler discovered that the ability to carry out the walk as described (or traverse the network) was based on analyzing vertices as being even or odd. **Odd vertices** have an odd number of paths going to them and **even vertices** have an even number of paths going to them. In the diagram above, vertex A is odd (5 paths), vertex B is odd (3 paths), vertex C is odd (3 paths), and vertex D is odd (3 paths). Euler proved that a network is traversable (traceable) if all of the vertices are even or if exactly two of the vertices are odd. Hence, the network pictured is not traversable.

Whenever a collection of things is joined by connectors, the mathematical model employed is a network or graph. A **network** is a figure or graph consisting of points (vertices or nodes) and edges (segments or arcs) that join various vertices to one another.

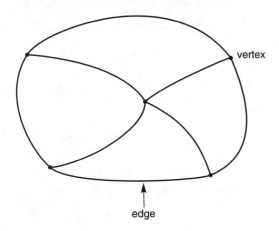

vertex

edge

 Model Problem

Explain why the following network is traversable. List a sequence that demonstrates a traceable path.

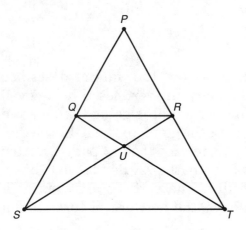

Solution   The numbers in parentheses indicate if a vertex is odd or even.

$$P(2) \quad Q(4) \quad R(4) \quad S(3) \quad T(3) \quad U(4)$$

Since exactly two of the vertices are odd, the network is traversable. The sequence

$$S \rightarrow T \rightarrow U \rightarrow Q \rightarrow R \rightarrow P$$

would be one traceable path.

1. For the network below, how many of the vertices are considered odd?

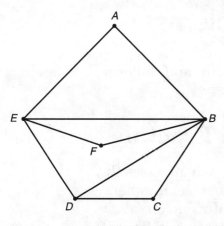

   A. 0   B. 2   C. 3   D. 4

2. Which of the following networks would NOT be traversable?

A.

B.
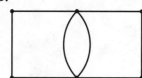

C.
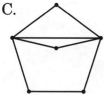

D.
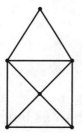

3. For the network below, how many of the vertices are even?

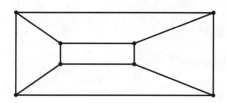

   A. 0   B. 2   C. 4   D. 8

4. A complete network has at least one path or edge between each pair of vertices. Which of the following are complete networks?

I.
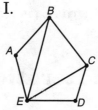
II.
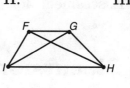
III.

A. I and III
B. II only
C. II and III
D. I, II and III

5. Draw a sketch of a network of five towns that would NOT be traversable.

6. A salesperson starts in Philadelphia (*P*) and must travel to Newark (*N*), Baltimore (*B*), and Scranton (*S*) before returning to Philadelphia. The distances between these cities are given. Find the shortest route the salesperson can take. Explain your approach. You may want to use a tree diagram to analyze the different routes.

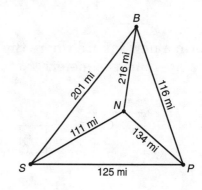

Use iterative patterns and processes to describe real-world situations and solve problems.

Note: The Assessment for Macro C is combined with and follows Macro D.

3 D 1 Recursion, Iteration, and Fractals

An **iteration,** or **recursion,** is a process in which each new result depends on the preceding results. Iterations may involve numbers or geometric figures. An iteration will contain an initial value (or *seed*) or geometric figure, a rule, and an output value or geometric figure, which then becomes the initial value for the next iteration of the rule.

Examples

a. a numerical iteration

Initial value: 5.
Rule: Multiply the previous value by 2.
Result: 5, 10, 20, 40, …

b. a geometric iteration

Start with a square.

Rule: Form a new square where the new perimeter is half as long as the previous perimeter.
Result:

Model Problem

1. An iterative process is used in which the rule is "Multiply by 5, and then add 2." If the seed value is 3, find the next four values obtained through the iterations.

Solution

> 3
> $5 \times 3 + 2 = 17$ (applying the rule)
> $5 \times 17 + 2 = 87$ (applying the rule a second time)
> $5 \times 87 + 2 = 437$ (applying the rule a third time)
> $5 \times 437 + 2 = 2{,}187$ (applying the rule a fourth time)

2. If Robin puts \$1,000 in a bank that gives 5.5% interest per year figured once a year, how much will she have in the bank at the end of five years? How is this considered an iteration?

Solution

> End of year 1: $1{,}000.00 + 0.055(1{,}000) = \$1{,}055.00$
> End of year 2: $1{,}055.00 + 0.055(1{,}055) = \$1{,}113.03$
> End of year 3: $1{,}113.03 + 0.055(1{,}113.03) = \$1{,}174.25$
> End of year 4: $1{,}174.25 + 0.055(1{,}174.25) = \$1{,}238.83$
> End of year 5: $1{,}238.83 + 0.055(1{,}238.83) = \$1{,}306.97$

Robin will have \$1,306.97 in the bank at the end of five years. This process is an iterative process since each calculation is the same as the previous calculation, except that the new calculation uses the result from the previous calculation as its starting value.

Compound Interest

Interest is a charge for money that is borrowed. When you take a loan from a bank, you pay the bank interest. When you deposit or invest money, you earn interest because the bank is using your money to conduct its business. **Principal** is the amount of money that is borrowed or invested.

The example of iteration in Model Problem 2 is known as compound interest. **Compound interest** is interest based on principal and previous interest. Compound interest allows you to earn interest on your interest. The compound interest formula is:

$$A = p(1 + r)^t,$$

> where A = amount with interest
> p = principal
> r = annual interest rate
> t = number of years

In applying this formula to Model Problem 2, we get:

$$A = 1,000(1 + 0.055)^5$$
$$A = \$1,306.97$$

Note: On a scientific calculator, use the $\boxed{y^x}$ key to enter the exponent value of *t*.

Fractals

A **fractal** is a figure obtained through iteration in which you can see *self-similarity*. That is, the closer you look at a fractal, the more you see the same image. A well-known fractal is the Sierpinski triangle. To construct the Sierpinski triangle, you start with an equilateral triangle as the initial figure and apply the iteration rule: remove from the middle of this triangle a smaller equilateral triangle whose side measures one-half of the original side length so that three congruent triangles remain.

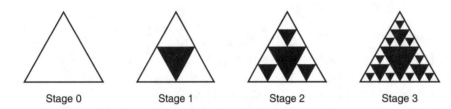

Stage 0   Stage 1   Stage 2   Stage 3

Observing the stages of a fractal, we see that the smaller and smaller details of the figure resemble the original, larger figure.

Model Problem

The diagram below shows stages 0, 1, and 2 for a fractal tree. Draw the next stage.

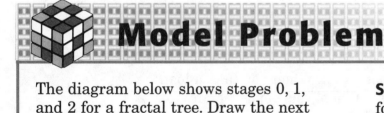

Stage 0   Stage 1   Stage 2

Solution To construct this fractal, for each straight branch you have to draw two branches one-half the original length one-third of the way down the branch at a 45° angle. Hence, stage 3 would look like:

1. The first three stages of a fractal are shown:

In the fourth stage of the fractal, how many circles will be of the smallest size?

A. 4   B. 8
C. 12   D. 15

2. An iterative process is used in which each term, after the first term, is found by adding 10 to 10 times the previous term. If the first term has a value of 5, what are the next three values obtained through the iteration?

3. A square has vertices at $(0, 0)$, $(9, 0)$, $(9, 9)$, and $(0, 9)$. A fractal is formed by dividing the sides of the square into three equal segments to form nine smaller squares, and then removing the middle square. Stage 1 is shown:

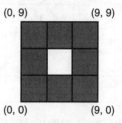

In stage 2, what are the coordinates of the vertices of the square removed that is closest to the origin?

Open-Ended Questions

4. Use the iteration rule $x \rightarrow x^2$ (to get a new value, square the current value) to investigate what happens for two different initial values: 2 and 0.5.

   a. Show the first five terms based on each initial value.
   b. What can you conclude about the limiting values that each sequence approaches?

5. Apply the following geometric iteration rule: Start with a unit square (side length of 1 unit). Split the square horizontally into two congruent rectangles. Remove the top rectangle, leaving behind the bottom rectangle.

   a. Show the first four stages of this iteration.
   b. What are the dimensions of the remaining rectangle in each stage?
   c. If we carry out the iterations indefinitely, do we generate a fractal? Explain your response.

6. To pay for acting classes, Julieta must borrow $800 from a bank. Her bank offers her a four-year loan with interest of 10% compounded annually.

a. Complete the table to show how the interest compounds. If Julieta accepts the bank's offer, how much will she owe at the end of four years?

Year	Principal	Interest	Total Owed
1			
2			
3			
4			

b. Apply the compound interest formula to find the answer to part a.
c. Julieta's uncle offers to lend her the $800 for four years with a simple interest rate of 20%. How does his offer compare with the bank's offer?

---

3 D 2 Algorithms and Flow Charts

An **algorithm** is a finite, step-by-step procedure for accomplishing a task. Familiar examples of algorithms include:

- doing long division
- computing percent increase
- writing the prime factorization of a composite number
- alphabetizing a list of words
- following a recipe

An **iteration diagram** is a pictorial representation of the algorithm associated with an iterative process, that is, a process involving repetition.

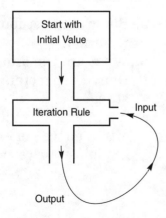

Model Problem

If eight iterations are performed based on the diagram shown, complete a table indicating an output value for each term.

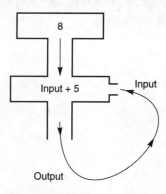

Solution

Term	0	1	2	3	4	5	6	7	8
Output	8	13	18	23	28	33	38	43	48

Note: The output value 8 is associated with term 0 since this value exists before the iterations.

A **flow chart** is a diagram illustrating a procedure or process. Flow charts are useful in illustrating complicated procedures involving multiple steps, repetition, and decision-making. The following flow chart illustrates the process of determining if a whole number is divisible by 3.

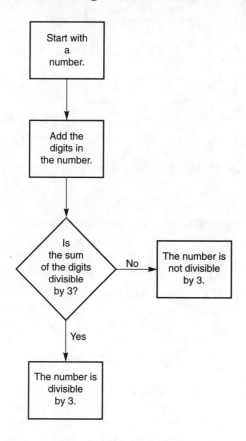

Model Problem

The Sieve of Eratosthenes is a procedure for identifying the prime numbers in a certain range of positive integers (from 1 to n). This algorithm is illustrated by the following flow chart:

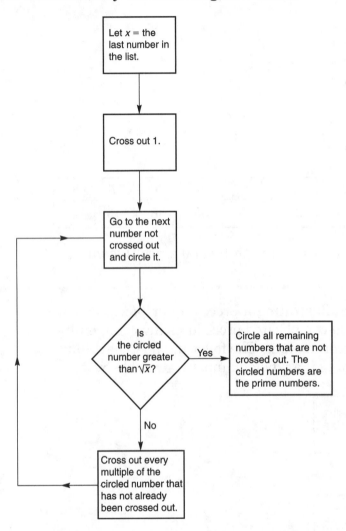

Use the algorithm to find the prime numbers from 1 to 50.

Solution

Follow the steps of the algorithm until you get to 8, since the first whole number greater than $\sqrt{50}$ is 8.

1̶	②	③	4̶	⑤	6̶	⑦	8̶	9̶	1̶0̶
⑪	1̶2̶	⑬	1̶4̶	1̶5̶	1̶6̶	⑰	1̶8̶	⑲	2̶0̶
2̶1̶	2̶2̶	㉓	2̶4̶	2̶5̶	2̶6̶	2̶7̶	2̶8̶	㉙	3̶0̶
㉛	3̶2̶	3̶3̶	3̶4̶	3̶5̶	3̶6̶	㊲	3̶8̶	3̶9̶	4̶0̶
㊶	4̶2̶	㊸	4̶4̶	4̶5̶	4̶6̶	㊼	4̶8̶	4̶9̶	5̶0̶

Answer The prime numbers from 1 to 50 are 2, 3, 5, 7, 11, 13, 17, 19, 23, 29, 31, 37, 41, 43, and 47.

1. Starting with $n = 10$, find the list of numbers produced by the following algorithm (for use with natural numbers greater than 1).

  Step 1: If n is odd, replace n by $7n - 5$.

  Step 2: If n is even, replace n by $\frac{n}{2}$.

  Step 3: If $n = 100$, stop; otherwise return to step 1.

In Questions 2 and 3, suppose that when one shape is inside another, you follow the inside direction first.

2. Given the following definitions:

 means $a + 4$

 means $a - 5$

 means $3a$

 means $4a$

Which of the following are true?

  I.

  II.

  III.

A. I only   B. II only
C. III only   D. II and III

3. Given the following definitions:

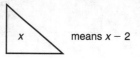 means $x - 2$

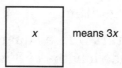 means $3x$

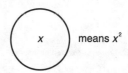 means x^2

Find the value of

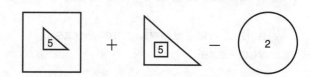

4. Write the expression indicated by the following flowchart.

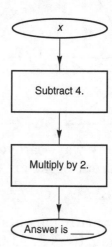

5. Describe the steps of an algorithm that will produce the following sequence of figures:

 , , ...

6. Produce a diagram resulting from the following algorithm:

Step 1: Start with a square having side lengths of 3 inches.

Step 2: Divide each side of the square into 3 equal segments.

Step 3: Remove or erase the middle segment of each side of the square.

Step 4: Construct a smaller square on the exterior of each side of the original square where the middle segment has been removed.

a. Draw the figure obtained by following the algorithm.

b. What is the perimeter of the final figure?

c. What is the area of the final figure?

Assessment Macro C and Macro D

1. Which of the following networks is traversable?

A.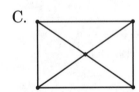

B.

C.

D.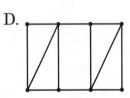

2. In how many ways can you arrange seven different books on a bookshelf?

A. 7^2 B. $7!$ C. $7\frac{1}{2}$ D. 7^7

3. How many arrangements are possible of all of the letters in ALGEBRA if each arrangement must begin and end with A?

A. $5!$ B. $6!$ C. $\dfrac{7!}{2}$ D. $7! - 2$

4. Given the iteration rule "Multiply by 2," what would be the value after the rule is applied five times starting with a seed value of 5?

A. 10   B. 40   C. 160   D. 640

5. Draw the next stage in the given fractal.

Stage 0   Stage 1

6. Using the digits 5, 6, 7, 8, with repetition of digits possible, how many four-digit numbers can be formed if the number must be greater than 8,000 and also a multiple of 5?

7. The digits 2, 3, 4 are used to form three-digit whole numbers, without any repetition of digits. What is the probability that a randomly selected number from among all those generated by the above is NOT an even number?

8. If four teams A, B, C, and D were playing in a single-elimination tournament, the following diagram would illustrate the progress of the tournament.

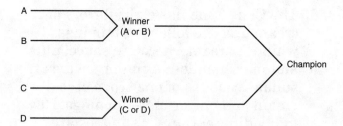

In this tournament, there were a total of three games. How many games would exist in a field of 64 teams for a single-elimination tournament?

9. A box fractal is formed by the following rule:

   Start with a square.

   Divide the square into 9 smaller squares.

   Remove 4 middle squares from each side.

   Continue the same procedure with each remaining square.

   Stages 0 and 1 are shown.

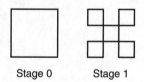

Stage 0   Stage 1

How many squares are there in stage 2?

10. Complete the rule that is defined by the operation ⋆ if

$$5 \star 2 = 27$$
$$3 \star 1 = 10$$
$$9 \star 4 = 85$$
$$10 \star 7 = 107$$
$$a \star b = \underline{\quad}$$

11. Many members of the student council are on more than one subcommittee, as shown in the table. An X indicates that these two subcommittees have at least one member in common.

If the subcommittees must all meet and the only time slot available each day is 2:30 P.M. to 3:30 P.M., what is the minimum number of days necessary for all of the subcommittees to meet?

	Special Events	Home-coming	Dances	Fund Raising	School Policy	Pub-licity
Special Events						X
Home-coming			X			
Dances		X		X	X	X
Fund-Raising			X			X
School Policy		X	X			X
Pub-licity	X		X	X	X	

12. Sketch the next 3 pictures using the iteration rule "Change the previous rectangle so that each figure is one half as long and twice as wide."

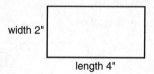

width 2"

length 4"

13. Starting at P and proceeding downward on a diagonal either left or right each time, determine how many different paths spell the word PRISM.

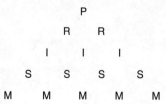

14. Find the starting number for the given flow chart.

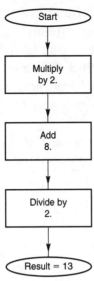

Start

Multiply by 2.

Add 8.

Divide by 2.

Result = 13

15. A map company is designing a map of the state of New Jersey. Each region of the state must be only one color. Regions that share a common boundary, except at a corner, may not be the same color. What is the LEAST number of colors needed?

16. The Cone Zone has eight flavors of ice cream and the following toppings: whipped cream, chocolate sauce, nuts, and marshmallow. A junior ice cream sundae consists of one scoop of ice cream and one or more toppings. How many different junior sundaes are possible?

Open-Ended Questions

17. The tree diagram shows all the possible outcomes for boys and girls in a family of three children.

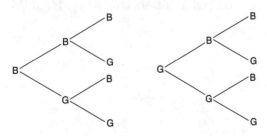

    a. Explain how the tree diagram shows that the probability of all boys is $\frac{1}{8}$.

    b. How would you expand this tree diagram to account for a family with four children?

    c. Which of the following situations could be represented by a tree diagram like the one shown? Explain your answer.

        (i) tossing a coin three times

        (ii) answering a three question multiple-choice quiz where each question has four possible answers

18. Suppose you were talking on the telephone to a friend. Write an algorithm you could read to your friend so that he or she could duplicate the given diagram without seeing it. Include any tools or materials you would ask your friend to use.

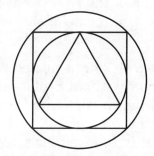

19. Use the iteration rule $x \to x^2 - 1$ to investigate what happens for two different initial values, 0 and -1.

   a. Show the first five terms based on each initial value.
   b. What can you conclude about the sequences formed by the iteration for each value?
   c. Name another initial value for which you would get a similar result.

20. A sorting algorithm can be used to arrange a list of n numbers in decreasing order. The algorithm makes successive passes through the list of numbers from left to right. In each pass, successive pairs of adjacent numbers are compared; if the number on the left is smaller than the one on the right, the two are exchanged; otherwise, they are left as is. Hence, after the first pass, the smallest number sorts all the way to the right.

Show how this sorting algorithm can be used to arrange the numbers 1, 7, 8, 2, 9, 6 in decreasing order. How many passes did it take until the numbers were in order?

21. The figure consists of 21 small congruent squares.

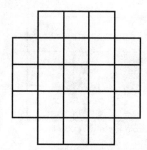

What is the total number of different squares (of any size) that one can trace using the lines of the figure? Explain the process you used to be sure that you counted all the squares.

ASSESSMENT Cluster 3

1. Franklin rolls two dice. What is the probability that he rolls a sum of either 3 or 8?

   A. $\frac{7}{36}$ B. $\frac{1}{6}$ C. $\frac{5}{36}$ D. $\frac{1}{12}$

2. Which of the following is TRUE about the following data?

Score	Frequency
80	2
82	4
86	3
90	5

   A. The mean is greater than the median.
   B. The median is greater than the mean.
   C. The mean equals the median.
   D. There is no mode for the data.

3. The whole numbers from 1 to 25 are each printed on separate slips of paper. You draw one slip of paper. Which of the following events would have a probability of 8%?

   A. You pick a multiple of 5.
   B. You pick a factor of 36.
   C. You pick an odd multiple of 7.
   D. You pick a multiple of 6.

4. The table shows the boxes of candy sold by each grade at Ocean Middle School.

Candy Sale Results	
Grade	**Boxes of Candy**
Fifth	11,200
Sixth	9,600
Seventh	11,700
Eighth	7,500

Which graph is the WORST representation of the data?

A.
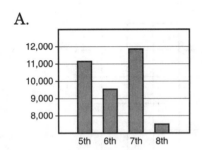

B.
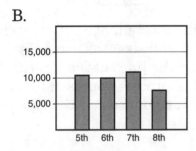

C.
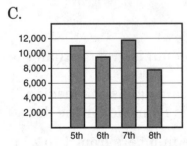

D.
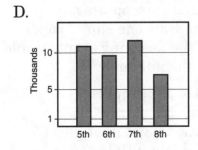

5. Which of the following situations represents *independent* events? (That is, the outcome of the first event has no effect on the outcome of the second event.)

A. Two marbles are selected from a container without replacement.
B. Two children born in a family are a boy and then a girl.
C. A ticket is selected for second prize after the first-prize ticket has been removed.
D. A king is selected from a deck of cards after two kings have already been dealt.

6. A theater did a survey of the ages of people who attended its cabaret. The table summarizes the results. Find the median age group of the attendees.

A. 13–30
B. 31–48
C. 49–66
D. over 66

Ages	Numbers
12 and under	87
13–30	298
31–48	481
49–66	364
over 66	95

7. The rules for a board game call for a player to lose a turn if the player rolls three consecutive doubles on a pair of dice. What is the probability that a player will lose a turn in her or his first three rolls?

A. $\frac{1}{6}$ B. $\frac{1}{18}$

C. $\frac{1}{36}$ D. $\frac{1}{216}$

8. Two dice are rolled. Latrell records the sum and Debra records the product. Who has a better chance of obtaining a 12?

A. Latrell
B. Debra
C. They have the same chance of obtaining a 12.
D. cannot be determined

9. How many arrangements are possible using all the letters of the word PRIME, if each arrangement must begin with an R?

A. 4!   B. 5!
C. 5! − 1   D. 4^4

10. Use the data in the table, which was found in the telephone directory, to determine which graph could represent the charges for calls.

Sample Telephone Day Rates		
From Passaic to	First Minute	Each Additional Minute
Freehold	0.33	0.11
Morristown	0.17	0.07
Newark	0.09	0.03
Toms River	0.37	0.11

A.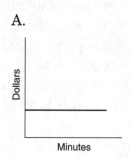

B.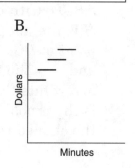

C.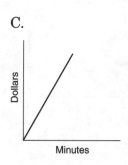

D.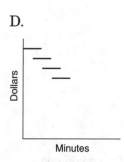

11. In how many different ways can you arrange all of the letters A, B, C, D, E, F, and G if each arrangement must begin and end with a vowel and the middle letter must be G?

A. 7!   B. 5!
C. 2 × 5!   D. 2 × 4!

12. Mr. Prior runs an appliance store. He purchases his appliances for a wholesale price. He marks the wholesale price up 80% to establish the selling price. If he were to use the spreadsheet shown to figure out the selling price, which formula should he enter into cell D2 to compute the selling price of the refrigerator?

	A	B	C	D
1	Item	Wholesale Price	Markup Rate	Selling Price
2	Refrigerator	$500	0.80	
3	Washer	$200	0.80	
4	Dryer	$300	0.80	

A. B2 × C2   B. B2 × C2 + B2
C. B2/C2   D. B2/C2 + B2

13. A contest offers a prize of a trip to London, Paris, or Rome in either spring, summer, or fall. How many different choices are possible if a prize consists of one city and one season?

14. If a pair of dice is rolled, what is the probability of getting a sum of 8?

15. The mean of a set of 10 scores is 61. What is the sum of the 10 scores?

16. Eight scores have a mean of 30. The bottom two scores have a mean of 21. The top two scores have a mean of 50. What is the mean of the four middle scores?

17. The probability of rain on Saturday is 60%. The probability of rain on Sunday is 30%. What is the probability that it will rain both days?

18. Starting at P and proceeding downward on a diagonal either left or right each time, determine how many different paths spell the word PASCAL.

```
            P
         A     A
      S     S     S
   C     C     C     C
 A     A     A     A     A
L     L     L     L     L     L
```

19. Which of the following is a better deal?

> $1,000 invested at 4% interest compounded annually for 6 years.
>
> $1,000 invested at 6% interest compounded annually for 4 years.

What is the difference between the two situations?

20. A survey of favorite pets was taken at the mall. 50 male shoppers and 50 female shoppers were surveyed. Based on the data in the table, if 500 people were surveyed, approximately how many would be expected to choose dogs as their favorite pet?

Favorite Pets

Pet	Male	Female
Bird	2	4
Cat	10	15
Dog	25	26
Fish	5	3
No pet	8	2

21. Morse code is a system of communicating information in which ordered sets of dots and dashes represent the letters of the alphabet, numerals, and punctuation. A single dot represents the letter *e*, while a dash followed by two dots represents the letter *d*. How many distinct letters can be represented by arrangements of 1 to 5 symbols, each of which is a dot or a dash?

22. A target has the shape of rectangle *ABCD*, with the dimensions shown. *H* is the midpoint of $\overline{AB}$. *E* is the midpoint of $\overline{DC}$. *F* is the midpoint of $\overline{EC}$, and *G* is the midpoint of $\overline{DE}$. If an arrow hits the target at random, what is the probability that it will land in the shaded region?

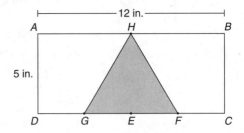

Open-Ended Questions

23. Give a set of data that has the following measures of central tendency:

   mode = 9
   mean = 12
   median = 12

Explain how you constructed the set.

24. The following data represent the heights in inches of students in a kindergarten class. Construct a line plot from the data. State the median, the mode, and the mean for the data.

   48 47 44 46 48 46 42 46 51
   46 50 43 42 45 43 47 49

25. Construct a bar graph from the information in this table.

Nicole's Fall Fund-Raiser Sales	
Gift Bags	ⅢⅠ ⅢⅠ ⅢⅠ ⅢⅠ ⅢⅠ ⅢⅠ Ⅲ
Rolls of Wrap	ⅢⅠ ⅢⅠ ⅢⅠ Ⅲ
Bags of Bows	ⅢⅠ ⅢⅠ Ⅲ
Gift Tags	ⅢⅠ ⅢⅠ Ⅰ

26. The Greek mathematician Euclid devised an algorithm for finding the greatest common factor of two numbers. The algorithm states:

Divide the greater number by the lesser number.
Divide the previous divisor by the remainder.
Repeat the process until the remainder is 0.

The greatest common factor is the divisor used in the last division. Show how Euclid's algorithm can be applied to find the greatest common factor for 38 and 98.

27. A discount clothing store has the following policy: Each week that an item remains on the rack it is discounted by 10% of its current price.

a. A suit is originally priced at $450. Generate a table to show the price of the suit for the first six weeks it is on the rack.
b. Write a formula to generalize the sequence of prices produced in part a.
c. Determine the number of weeks it will take for the suit to be priced less than $100.

28. The gymnastics team competes in five events. The team members have signed up as follows for the events.

Floor Exercises: Jeanne, Michelle
Horse: Jeanne, Dora
Parallel Bars: Jeanne, Michelle, Olga, Dora
Balance Beam: Michelle
Rings: Olga, Dora

Every practice has three time blocks, during which the coach supervises two different events. Design a practice schedule that lets each girl practice all her events and that avoids any conflicts.

29. Two dice are rolled.

a. Explain why the probability of obtaining a sum less than or equal to 5 is the same as the probability of obtaining a sum greater than or equal to 9.
b. If the probability of obtaining a sum less than or equal to 6 is the same as the probability of obtaining a sum greater than or equal to k, find the value of k.
c. If three dice are rolled, what is the probability that the sum obtained is less than or equal to 3?
d. The probability that the sum obtained from three dice is greater than or equal to x is the same as the answer to part c. What is the value of x

30. Describe in writing the process used to generate the fractal shown. Be sure to indicate where the process repeats. Then draw the fractal as it would appear in stage 3.

Stage 0   Stage 1   Stage 2

31. Use the data in the table for questions a–c.

U.S. Population	
Year	Population (in millions)
1890	63
1900	76
1910	92
1920	106
1930	123
1940	132
1950	151
1960	179
1970	203
1980	227
1990	249
2000	281

a. Draw a line graph of the data given in the table.
b. Using the graph, find the U.S. population in 1975.
c. Examine the population changes between 1980 and 1990 and between 1990 and 2000. Predict what the population might be in the year 2010 if this pattern continues. Explain your reasoning.

32. A bicycle race is set up so that a cyclist must visit each station on the course to have his or her card stamped. Stations can be visited in any order. The course is pictured below. The letters represent the sta-tions and the numbers represent the distances (km) between stations.

a. Find a route through the course that enables a cyclist to visit each station and pass through it exactly once.
b. Find the shortest route that includes each station at least once.
c. Compare the conditions of parts a and b of this question.

33. Construct a scatter plot using the following information. Is there a correlation? If so, is it positive or negative? What can you conclude about time spent doing homework and time spent watching television?

Time Spent Doing Homework vs. Watching Television		
Student	Homework (min)	Television (min)
A	30	60
B	90	45
C	90	0
D	75	90
E	60	120
F	75	30
G	60	60
H	60	0
I	0	180
J	45	30

34. A restaurant offers five entree choices on its dinner menu: pasta, hamburger, chicken, stir-fry, and fish. With each entree, the customer may choose one type of potato: French fries, mashed, or baked, and one of two desserts: ice cream or pudding.

a. Make a tree diagram to show the possible meal choices.
b. How many different meals are available assuming the customer selects an entree, potato, and dessert?
c. Suppose the diner has the option of not taking a potato or a dessert. How does that change the number of different meals that are available? Show all work.

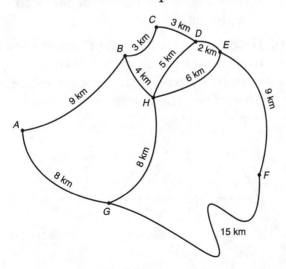

CUMULATIVE ASSESSMENT

Clusters 1, 2, and 3

1. How many three-digit numbers have all of the following characteristics?

   I. The number is a multiple of 72.
   II. The number is divisible by 5.
   III. The number is less than 500.

   A. 0   B. 1
   C. 2   D. 5

2. The diameter of the circle is 10 inches. What is the length of arc *AB*, if the central angle *AOB* is 120°? (Give the answer to the nearest tenth of an inch.)

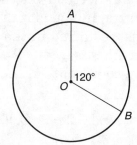

   A. 5.2 inches   B. 10.5 inches
   C. 15.7 inches   D. 26.2 inches

3. Which of the following is NOT equal to the other numbers?

   A. 0.2   B. $\sqrt{\dfrac{1}{25}}$

   C. $\dfrac{1}{2} + \dfrac{1}{3}$ D. $\dfrac{0.05}{0.25}$

4. Triangle *ABC* is isosceles with $\overline{AB} \cong \overline{AC}$ and

   $$\angle DAB \cong \angle BAC \cong \angle EAC.$$

   What is the measure of $\angle EAC$?

   A. 40   B. 55
   C. 70   D. 80

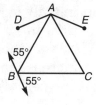

5. The Cultural Studies club consists of only seventh and eighth graders. If the ratio of seventh graders to eighth graders in the club is 3:2, which of the following could NOT be the total number of club members?

   A. 16   B. 20   C. 25   D. 30

6. Using four of the five digits 1, 3, 4, 7, 8, how many four-digit numbers can be formed, with no repetition of digits, if the number must be odd?

   A. 625   B. 120   C. 72   D. 36

7. Which of the following does NOT exist?

   A. a rhombus with two angles each measuring 5°
   B. a quadrilateral with one angle measuring 80° and each additional angle measuring 10° more than the previous one
   C. a trapezoid with two right angles
   D. an obtuse isosceles triangle

8. Neela rolls three dice. What is the probability that she will obtain a sum that is less than or equal to 4?

   A. $\dfrac{1}{216}$ B. $\dfrac{1}{108}$ C. $\dfrac{1}{72}$ D. $\dfrac{1}{54}$

9. The line plot shows test scores for fifteen students:

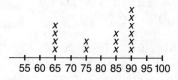

   Which of the following is true about the data?

   A. The mean equals the median.
   B. The median is greater than the mean.
   C. The median is less than the mean.
   D. There is no mode for the data.

10. During a recent month, the exchange rate of Canadian dollars to U.S. dollars was 1 to 0.81. If you paid $65 in Canadian dollars for a toaster oven, what would you have paid in U.S. dollars? (Disregard tax.)

A. $52.65 B. $65.81
C. $80.25 D. $84.00

11. *ABCD* is a square with side length of 4 cm. $\overline{AP}$ and $\overline{AR}$ are congruent line segments each 1 cm long. Approximately what percent of the area of the square is the area of the triangle *RAP*?

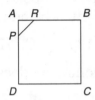

A. 3% B. 6% C. 10% D. 12%

12. About what percent of the big square is shaded?

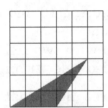

A. 5% B. 15% C. 20% D. 25%

13. Which of the following products would have the greatest absolute value?

A. (6,000,000)(0.006)
B. (−400)(−300)(−200)
C. $(2.5 \times 10^2)(2.5 \times 10^{-2})$
D. (5,000)(−5,000)

14. There are four times as many boys as girls on the newspaper staff of Beaverton Middle School. If there are 40 staff members in all, how many of them are girls?

A. 8 B. 10 C. 30 D. 32

15. Point *A* is reflected over the *y*-axis and then the image is reflected over the *x*-axis, resulting in the point *A″* with coordinates (−4, −2). What was the *y*-coordinate of the original point *A*?

A. −4 B. −2 C. 0 D. 2

16. Yoshi bought a sweater at 30% off the original price. The discount saved him $12.60. What was the original price of the sweater?

A. $8.82 B. $21.42
C. $29.40 D. $42.00

17. Which of the following would be a reasonable value for the percent of the figure that is shaded?

A. 75% B. 40% C. 20% D. 10%

18. The population in Culver Heights increased from 1990 to 1993 as shown in the table. What was the average annual percent increase in population over the three-year period?

Years	Population
1990 to 1991	10,000 to 11,000
1991 to 1992	11,000 to 12,000
1992 to 1993	12,000 to 13,000

A. 8.33% B. 9.14%
C. 10% D. 27.42%

19. A square has vertices at $(-3, 0)$, $(0, 3)$, $(3, 0)$, and $(0, -3)$. How many of the following points are in the exterior of the square?

$(-1, 1)$ $(-2, -2)$ $(0, -4)$
$(2, 2)$ $(1.5, 1.5)$ $(-3, 1)$

A. 2 B. 3 C. 4 D. 5

20. Set G consists of the three-digit multiples of 3.
Set H consists of the three-digit multiples of 4.
Set L consists of the three-digit multiples of 6.
Which of the following statements is true about the three sets of numbers?

A. If a number is in set G, it is also in set L.
B. The smallest number contained in all three sets is 144.
C. If a number is in set L, it is also in set G.
D. No multiple of 9 is in set H.

21. There are 6 members on the student council executive committee. In how many ways can a 2-member subcommittee be formed?

22. There are 30 students in a class. The mean grade for that class on a test was computed as 80, but two students' grades were read incorrectly as 90 instead of 50. What will the average grade (rounded to the nearest tenth) be when it is recomputed using the correct scores?

23. How many different isosceles triangles are possible where the side-lengths are whole numbers and the perimeter is 31 units?

24. The diagram shows a structure made with nineteen cubes. If each cube has an edge of 1 cm, what is the surface area of the structure (in square centimeters)?

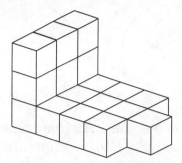

25. Figure A was enlarged and flipped to make figure B. Find the ratio of the side lengths of the two figures.

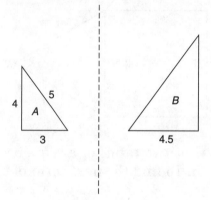

26. Winter Spring bottled water comes in three sizes:

6-pack of 0.5-liter bottles at $3.59 per 6-pack

1-liter bottles at $0.99 each

1.5-liter bottles at $1.59 each

If you need 9 liters of bottled water, how much would you save by buying 1.5-liter bottles instead of 6-packs of 0.5-liter bottles?

27. Evaluate the expression:

$14 - (-2)^3 + 6 \div (-3)$.

28. The New York Yankees plan to raise the price of bleacher seats from $15 to $17.50. What is the percent of increase?

29. The two spinners shown have eight and four congruent sections, respectively. If you spin each spinner once, what is the probability of obtaining the largest possible sum?

30. *ACEF* is a square. *ABC* is an isosceles right triangle. *CDE* is an equilateral triangle. Find the measure of ∠*BCD*.

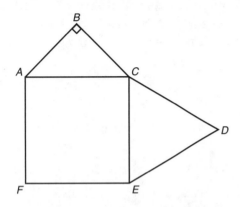

31. The first two terms of a sequence are shown. To find the next term of the sequence, find the area of a square whose sides are twice as long as the sides of the previous square. What is the area of the sixth term?

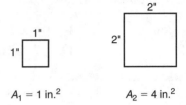

$A_1 = 1$ in.² $A_2 = 4$ in.²

32. Arrange the following numbers from LEAST to GREATEST.

1.34×10^{-3} 1.3×10^{-2} -2.5×10^{-1}

33. Imagine you have a collection of congruent equilateral triangles. What is the minimum number of these equilateral triangles that you can place together to form the following figures?

a. a rhombus
b. an isosceles trapezoid
c. a regular hexagon

34. How many cubes with an edge of 4 cm would give you the same volume as a rectangular prism with dimensions 16 cm by 8 cm by 5 cm?

Open-Ended Questions

35. These are test scores in Madame Dubin's French class:

93, 92, 84, 81, 68, 81, 78, 77, 84, 63, 62, 90

a. Construct a line plot for the data.
b. State the median score for the data.
c. What would happen to the median if one point were added to each of the test scores? Explain.

36. A package of gum has a price increase from $0.50 to $0.60. At the old price, a package contained 10 sticks of gum; now a package contains 8 sticks.

Determine the percent increase in going from the old situation to the new. Explain your procedure and thinking.

37. Find the area and perimeter of the hexagon. Show your complete procedure.

38. The School Supplies Megastore offers two brands of pencils:

Brand A: a dozen pencils for $2.35
Brand B: 18 pencils for $3.75

    a. Which is a better buy? Show how you arrived at your answer.

    b. Suppose Brand C has 15 pencils in a package. What would be a reasonable cost for the package so that the price is comparable to the offers of Brand A and Brand B? Explain your answer and show how you determined the answer.

39. Kathryn invests $2,000.00 for six years at 10% interest compounded annually. After the first three years, Elizabeth invests $3,000.00 for three years at 5% interest compounded annually. At the end of the six years, who would have more money? Explain.

40. Given the following four views for a three-dimensional figure, draw the figure on isometric paper (shown). Then give the volume and surface area for the figure drawn.

Left View   Front View   Right View   Top View

Extra Practice

Open-Ended Questions

Study Questions 1 and 2, which are presented with their solutions, and then try Questions 3 and 4.

When responding to an open-ended question, think about what you must do to form a response that will receive a score of 3.

- Answer all parts of the question.
- Present your work clearly, so that the person grading it will understand your thinking.
- Show all your work, including calculations, diagrams, and written explanations.

1. Katrina has the following five grades on her vocabulary tests in Spanish.

    82   98   84   98   92

- Katrina's teacher will use either the mean of the test grades or the median of the test grades to determine Katrina's average. Which would be better for Katrina? Explain why.
- Katrina is about to take the sixth vocabulary test. She wants to bring her mean score for all six tests up to 92. What does she need to score on this test in order to make her mean become 92? Show how you arrived at your answer.

Solution (for a score of 3)

Mean: The sum of the five grades is 454. The mean is $454 \div 5 = 90.8$

Median: Write the five grades in order to find the median.

$$82 \quad 84 \quad 92 \quad 98 \quad 98$$

The median is 92.

Since the median is greater than the mean, it would be better for Katrina if the teacher uses the median of 92.

The sum of the first 5 grades is 454.

If the mean of the 6 grades is to be 92, then the sum of the 6 grades would be

$$6(92) = 552.$$

To find out how many points Katrina needs, find the difference between what she has after 5 tests and what she needs to have after the sixth test.

$$454 + x = 552$$
$$x = 552 - 454$$
$$x = 98$$

To bring the mean up to 92, Katrina must score a 98 on the sixth test.

2. Omar is playing a board game that uses a pair of cubes with faces numbered 1, 1, 2, 2, 3, 3. To find how many spaces he can move on the board, he adds the two numbers he rolls.

- What are all of the possible sums Omar can roll?
- Is he more likely to get a sum of 2 or a sum of 4? Or are they equally likely? Explain your reasoning in detail.
- Omar needs to roll a sum of 5 in order to get another turn. What is the probability that he will roll a sum of 5? Explain your reasoning in detail.

Solution (for a score of 3)

$1 + 1 = 2$	$1 + 3 = 4$	$2 + 3 = 5$
$1 + 2 = 3$	$3 + 1 = 4$	$3 + 2 = 5$
$2 + 1 = 3$	$2 + 2 = 4$	$3 + 3 = 6$

The list shows all the sums possible by rolling the two dice. The possible sums are 2, 3, 4, 5, and 6.

A sum of 2 can only occur with a 1 on each number cube. From the list of sums, you can see that the probability of getting a sum of 2 is $\frac{1}{9}$.

A sum of 4 can occur with a 2 on each cube or a 1 on one cube and a 3 on the other. This makes the sum of 4 more likely. From the list of sums, you can see that the probability of getting a sum of 4 is $\frac{3}{9}$, which equals $\frac{1}{3}$.

Since $\frac{1}{3} > \frac{1}{9}$, the probability of getting a 4 is greater.

In order to get a sum of 5, Omar must get a 2 on the first cube and a 3 on the second, or the reverse.

$$P(2, 3) = \left(\frac{1}{3}\right)\left(\frac{1}{3}\right) = \frac{1}{9}$$

The probability of the reverse is also $\frac{1}{9}$. Therefore, the probability of obtaining a sum of 5 is $\frac{2}{9}$.

3. A set of data consists of the values 27, 33, 29, 27, x, and 35. Find a possible value of x so that

- there is one mode
- there are two modes
- the median is 28
- the mean is 35

4. Two ordinary dice are tossed and the sum of the numbers showing is found.

- What is the probability of getting a 5?
- Which is more likely, a sum of 7 or a sum of 8? Explain your answer.
- Which other sum has the same probability as a sum of 12?

Patterns, Functions, and Algebra

Recognize, create, and extend a variety of patterns and use inductive reasoning to understand and represent mathematical and other real-world phenomena.

4 A 1 Patterns

In a number of applications, problems can be solved by discovering a pattern and using the pattern to draw different conclusions. The pattern might be numerical or visual.

Computational Patterns

Many numerical situations involve patterns. To discover a numerical pattern, consider how each term is related to the terms before and after it.

Examples Observe the numerical patterns in each case.

a. repeating decimal

$$\frac{1}{11} = 0.0909\ldots$$

b. equivalent fractions

$$\frac{1}{2} = \frac{2}{4} = \frac{3}{6} = \frac{4}{8}$$

c. powers of ten

$$10^1 = 10$$
$$10^2 = 100$$
$$10^3 = 1,000$$

$$\bullet\bullet\bullet$$

$$10^9 = 1,000,000,000$$

d. concept of percent

$$1\% = \frac{1}{100}$$
$$10\% = \frac{10}{100}$$
$$50\% = \frac{50}{100}$$
$$100\% = \frac{100}{100}$$

e. multiplication pattern

$$9 \times 1 = 9$$
$$9 \times 2 = 18$$
$$9 \times 3 = 27$$
$$9 \times 4 = 36$$
$$9 \times 5 = 45$$
$$9 \times 6 = 54$$
$$9 \times 7 = 63$$
$$9 \times 8 = 72$$
$$9 \times 9 = 81$$

Model Problem

1. What is the units digit in the number equivalent of 3^{24}?

Solution Since it is not convenient to compute the value of 3^{24}, you need to see if a pattern exists by examining small powers of 3.

$$\begin{aligned}3^1 &= 3\\3^2 &= 9\\3^3 &= 27\\3^4 &= 81\end{aligned}\Bigg\}$$ For the first four powers of 3, the units digits are different (namely, 3, 9, 7, 1).

Then the units digit begins to repeat.

$$3^5 = 243$$
$$3^6 = 729$$
$$3^7 = \ldots7$$
$$3^8 = \ldots1$$

Assume that the pattern for the units digit continues as 3, 9, 7, 1. Note that this pattern is in groups of 4. You may conclude that every power of 3 that is a multiple of 4 will have a units digit of 1.

Answer Since 3^{24} is a power of 3 that is a multiple of 4, the value of 3^{24} will have a units digit of 1.

2. In the decimal representation for $\frac{5}{33}$, what digit would be in the 30th decimal place?

Solution The decimal representation for $\frac{5}{33}$ is 0.15151515.... Notice that in the repeating pattern, a 1 is in every odd position and a 5 is in every even position.

Answer The digit in the 30th decimal place would be a 5.

Visual Patterns

Visual patterns often appear in flooring, wallpaper, wrapping paper, store displays, and fabric. Many problem-solving situations involve understanding and extending visual patterns.

Example

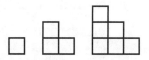

We can see that to generate the next term in this pattern, we must add a row of squares one greater in length than the bottom row of the previous term.

To Extend a Visual Pattern:

- Draw the pattern and try to figure out what has changed from one figure to the next.
- Describe the pattern in words.
- Convert the visual pattern to a numerical pattern.
- Use a table to organize your thinking.

 Model Problem

1. How many unit squares are needed to represent the fifth term of the following pattern?

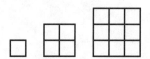

Solution

Method 1 Observe and draw the pattern.

Notice that the first term is a 1 × 1 square, the second term is a 2 × 2 square, and the third term is a 3 × 3 square. Then, draw the fourth term (4 × 4) and the fifth term (5 × 5).

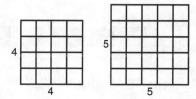

Counting the unit squares in the fifth term produces a result of 25 unit squares.

Method 2 Convert to a numerical pattern.

Counting the number of unit squares used to build the three given terms produces the numerical pattern 1, 4, 9. Since these numbers are consecutive perfect squares (1^2, 2^2, 3^2), you realize that the fifth term would have to be the fifth square, 5^2, or 25.

Answer 25 unit squares

2. a. Draw the fourth figure in the pattern.

   b. What fraction of the whole square is represented by the *smallest* shaded square in the figure you drew for part a?

Solution

a.

b. Start with the first figure and see if a numerical pattern emerges.

No part of the first figure is shaded.

$\frac{1}{4}$ of the second figure is shaded.

The smaller shaded square in the third figure is $\frac{1}{4}$ of $\frac{1}{4}$, which is $\frac{1}{16}$.

The smallest shaded square in the fourth figure is $\frac{1}{4}$ of $\frac{1}{16}$, which is $\frac{1}{64}$.

Answer $\frac{1}{64}$

3. Observe the pattern

   YUMMYYUMMYYUMMY…

If the pattern is continued, what letter will be in the 105th position?

Solution Notice that the pattern repeats every five terms. Also, notice that every fifth term is Y.

Term	Letter
1st	Y
2nd	U
3rd	M
4th	M
5th	Y
10th	Y
15th	Y

Answer Since 105 is a multiple of 5, Y will be the 105th letter.

PRACTICE

1. What is the units digit in 2^{40}?

   A. 2   B. 4
   C. 6   D. 8

2. When raised to any whole-number power, which of the following numbers does NOT always produce the same units digit?

   A. 10   B. 9   C. 6   D. 5

3. Which of the following fractions converts to a repeating decimal?

   A. $\frac{1}{6}$ B. $\frac{1}{4}$

   C. $\frac{3}{16}$ D. $\frac{1}{25}$

4. What digit is in the 45th decimal place in the decimal value of $\frac{7}{11}$?

A. 1   B. 3
C. 6   D. 7

5. George decides he is going to start saving pennies in a large, empty pickle jar. On Monday, he puts 1 cent into the jar. On Tuesday, he doubles the amount to 2 cents. On each succeeding day, he doubles the number of pennies he put in the day before. How many days will it take George to save at least $20?

A. 11   B. 12
C. 15   D. 26

6. Which of the following decimals shows a pattern equivalent to the visual pattern below?

⊢ T ⊣ ⊥ ⊢ T ⊣ ⊥

A. 0.12891289   B. 0.313131
C. 0.541541541   D. 0.7777

7. Analyze the pattern:

PENCILPENCILPENCIL…

If the pattern is continued, what letter will be in the 83rd position?

A. P   B. E
C. I   D. L

8. Juan and Kim started a debating club. In September, they were the only members. They decided that once a month each member would debate every member of the club. They also decided to add one new member to the club each month. Which of the following patterns could be used to determine the total number of debates in the fifth month the club was operating?

A. 2, 6, 12, 20, 30
B. 1, 4, 9, 16, 25
C. 1, 3, 6, 10, 15
D. 2, 3, 4, 5, 6

9. If the pattern below continues until all letters of the alphabet are shown, how many letters, including repetitions, will precede the last Z?

ABBCCCDDDD

10. How many dots are needed to represent the first five terms, in total, for the given sequence?

11. Given that: $\frac{1}{2} + \frac{1}{4} = \frac{3}{4}$

$\frac{1}{2} + \frac{1}{4} + \frac{1}{8} = \frac{7}{8}$

$\frac{1}{2} + \frac{1}{4} + \frac{1}{8} + \frac{1}{16} = \frac{15}{16}$

Find: $\frac{1}{2} + \frac{1}{4} + \frac{1}{8} + \frac{1}{16} + \frac{1}{32}$.

12. The Silver Diner has small tables that seat 4 people, one on each side. When the restaurant must seat larger groups of people, tables are put together so that they share a common side. When 2 tables are put together, 6 people can be seated. If 5 tables are put together into a long row, how many people can be seated?

13. Jared painted a 4 × 4 × 4 cube green on all 6 faces. When the paint dried, Jared cut the cube into 64 smaller cubes (1 × 1 × 1). If Jared looked at each small cube, how many would have green paint on exactly 2 faces? In completing this problem, discuss cases involving a smaller original cube in order to show a pattern to use to answer the question.

14. Mary notices that on a 2 × 2 checkerboard there are 5 squares of various sizes.

 This 2 × 1 board has four 1 × 1 squares and one 2 × 2 square.

Mary thinks that a 4 × 4 checkerboard would have twice as many squares of different sizes. Do you agree or disagree with Mary's idea? Explain your reasoning.

15. For the given sequence, determine the total number of squares needed to represent the fifth term.

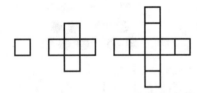

16. If the pattern is continued, how many dots would be in the 20th diagram?

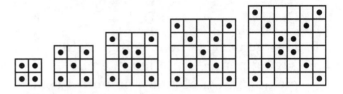

17. In successive stages of the pattern shown, each side of the equilateral triangle has a length equal to 110% of the length in the previous stage. To the nearest hundredth, what would be the perimeter of the triangle in the 6th stage?

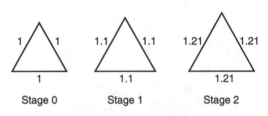

18. Suppose this pattern is continued. On the eighth figure, what percent of the figure is shaded? Explain your thought process.

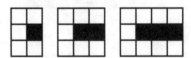

---

4 A 2 Sequences

A **sequence** is a list of numbers in a particular order that follows a pattern.

Arithmetic Sequences

An **arithmetic sequence** has a **common difference** between two consecutive terms.

Example

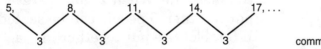

common difference = 3

Notice that the numbers are increasing. Notice also that they are increasing by 3. The next term is 17 + 3 = 20.

Model Problem

1. Which of these are arithmetic sequences?

    I. 9, 15, 21, 27, 33,…
   II. 18, 10, 2, −6, −14,…
  III. 7, 11, 16, 22, 29,…

Solution   Check for a common difference.

I.
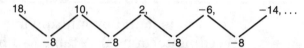
   common difference

II.
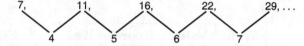
   common difference

III.

  7,   11,   16,   22,   29, …

    4   5   6   7   no common difference

Answer   I and II are arithmetic sequences. While III does follow a pattern, it is not an arithmetic sequence because the difference keeps increasing.

2. Find the 50th term of the arithmetic sequence:

$$3, 7, 11, 15, 19, 23,…$$

Solution   In the sequence, the 2nd term is found by adding the common difference of 4 to the 1st term. Each successive term is found by adding 4 to the preceding term. A table can help reveal a pattern.

Term	Value	Expressed as
1st	3	3 + 0(4)
2nd	7	3 + 1(4)
3rd	11	3 + 2(4)
4th	15	3 + 3(4)
5th	19	3 + 4(4)
6th	23	3 + 5(4)

To get the 2nd term, add one 4. To get the third term, add two 4's, and so on. Hence, to get the 50th term, add forty-nine 4's.

$$3 + 49(4) = 199$$

Answer   The 50th term is 199.

Geometric Sequences

A **geometric sequence** has a **common ratio** between two consecutive terms.

Example   2, 6, 18, 54, 162,...

$$\frac{6}{2} = 3 \qquad \frac{18}{6} = 3 \qquad \frac{54}{18} = 3 \qquad \frac{162}{54} = 3$$

There is a common ratio of 3. The next term is $162 \times 3 = 486$.

 Model Problem

1. Which of these are geometric sequences?

    I. 16, 8, 4, 2, 1,...
    II. 10, 20, 80, 640,...
    III. 10, 50, 250, 1,250,...

Solution   Check for a common ratio.

I. 16, 8, 4, 2, 1,...

$$\frac{8}{16} = \frac{1}{2} \qquad \frac{4}{8} = \frac{1}{2} \qquad \frac{2}{4} = \frac{1}{2}$$

There is a common ratio of $\frac{1}{2}$.

II. 10, 20, 80, 640,...

$$\frac{20}{10} = 2 \qquad \frac{80}{20} = 4$$

There is no common ratio.

III. 10, 50, 250, 1,250,...

$$\frac{50}{10} = 5 \qquad \frac{250}{50} = 5 \qquad \frac{1,250}{250} = 5$$

There is a common ratio of 5.

Answer   I and III are geometric sequences.

2. Find the 10th term of the geometric sequence:

$$\frac{1}{2}, 1, 2, 4, 8,...$$

Solution   The common ratio is 2. The 2nd term is found by multiplying $\frac{1}{2}$ by the common ratio 2. Then each successive term is found by multiplying the preceding term by 2. A table can help reveal a pattern.

Term	Value	Expressed as	Equals
1st	$\frac{1}{2}$	$\frac{1}{2}$	$\frac{1}{2} \times 2^0$
2nd	1	$\frac{1}{2} \times 2$	$\frac{1}{2} \times 2^1$
3rd	2	$\frac{1}{2} \times 2 \times 2$	$\frac{1}{2} \times 2^2$
4th	4	$\frac{1}{2} \times 2 \times 2 \times 2$	$\frac{1}{2} \times 2^3$
5th	8	$\frac{1}{2} \times 2 \times 2 \times 2 \times 2$	$\frac{1}{2} \times 2^5$

To get the 2nd term, multiply by the 1st power of 2.

To get the 3rd term, multiply by the 2nd power of 2.

To get the 4th term, multiply by the 3rd power of 2, and so on.

Hence, to get the 10th term, multiply by the 9th power of 2.

$$\frac{1}{2} \times 2^9 = \frac{1}{2}(512) = 256$$

Answer   The 10th term is 256.

Note: 2^9 can be evaluated using the $\boxed{y^x}$ key on a calculator.

Fibonacci Sequence

Not every sequence is arithmetic or geometric. Some sequences are formed by specific rules. For example, the **Fibonacci sequence** is generated by adding the two previous terms to form the next term.

1,   1,   2,   3,   5,   8,...

(1 + 1)   (1 + 2)   (2 + 3)   (3 + 5)

 Model Problem

In the Fibonacci sequence 1, 1, 2, 3, 5, 8,...

a. What is the 10th term of the sequence?
b. What is the sum of the first 10 terms?

Solution

a. To determine the 10th term, extend the sequence using the rule "add the two previous terms to form the next term."

Answer 1, 1, 2, 3, 5, 8, 13, 21, 34, 55

b. To find the sum, add the 10 given numbers.

$$1 + 1 + 2 + 3 + 5 + 8 + 13 + 21 + 34 + 55 = 143$$

Answer The sum of the 10 terms is 143.

 PRACTICE

1. Which of the following is a geometric sequence?

   A. $6, 7\frac{1}{3}, 8\frac{2}{3}, 9,...$

   B. $\frac{1}{2}, \frac{1}{3}, \frac{1}{4}, \frac{1}{5},...$

   C. $-10, -100, -1,000,...$

   D. $2, 4, 2, 4, 2, 4,...$

2. Which of the following would give you the 20th term of the arithmetic sequence 6, 13, 20, 27,...?

   A. 20×6 B. $6 + 20 \times 7$
   C. 20×7 D. $6 + 19 \times 7$

3. Which of the following is NOT an arithmetic sequence?

   A. $2, 8, 32, 128,...$

   B. $\frac{1}{2}, \frac{3}{4}, 1, \frac{5}{4},...$

   C. $10, 10, 10, 10,...$

   D. $8, 4, 0, -4,...$

4. Which of the following would NOT be a term of this geometric sequence?

   $$3, 6, 12, 24,...$$

   A. 48   B. 64   C. 96   D. 192

5. Which is the next term in the given Fibonacci sequence?

$$1, 1, 2, 3, 5, 8, \ldots$$

A. 11   B. 13   C. 16   D. 40

6. Which of these sequences is a Fibonacci sequence?

A. $1, 4, 5, 9, 14, 23, \ldots$
B. $1, 4, 4, 16, 64, \ldots$
C. $1, 4, 9, 16, 25, \ldots$
D. $1, 4, 8, 12, 16, \ldots$

7. In the geometric sequence $3, -9, 27, -81, \ldots$, what is the sixth term?

A. -729 B. -243
C. 243 D. 729

8. Chairs are set up in a triangular arrangement with 1 chair in the first row, 3 chairs in the second row, 5 chairs in the third row, and so on. If the pattern continues, how many chairs would be in row 20?

9. A special sequence is formed by taking twice the sum of the two previous terms to find the third term and all succeeding terms. If the first four terms are 1, 2, 6, and 16, find the 8th term.

10. In an arithmetic sequence, the 5th term is 23 and the 7th term is 33. Find the common difference for the sequence.

11. The first four terms of an arithmetic sequence are 2, 8, 14, 20, and 122 is the 21st term. What is the value of the 20th term?

12. In this geometric sequence, what is the common ratio?

$$81, 27, 9, 3, \ldots$$

13. If the 9th term of an arithmetic sequence is 100 and the 10th term is 111, find the value of the first term.

14. Create an arithmetic sequence of at least six terms for which the common difference is -3. Explain why the sequence you wrote is arithmetic. Also, explain why there would be an infinite number of possible sequences fitting the given condition.

15. Jeannie's Candy Shop had sales of $50,000 during its first year of operation. If the sales increase by $6,000 per year, what will be the total sales in the eleventh year?

16. Consider the sequence $\frac{1}{2}, \frac{1}{4}, \frac{1}{8}, \frac{1}{16}, \ldots$

a. Give the next three terms in the sequence.
b. Is the sequence arithmetic, geometric, or neither? Explain.

---

4 A 3 Representations of Relationships and Patterns

The relationship between the perimeter of a square and the length of its sides can be expressed in a variety of ways.

a. Verbal statement: "The perimeter of a square is four times the length of a side."

b. Table of values:

Side Length	Perimeter
1	4
2	8
3	12
4	16

c. Set of ordered pairs: {(1, 4), (2, 8), (3, 12), (4, 16)}

d. Equation: $P = 4s$, where P is the perimeter, and s is the length of a side of the square.

e. Graph: Plot the ordered pairs to obtain the graph.

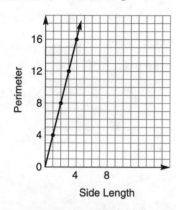

Model Problem

1. A can of soda costs $0.75. The amount of money you spend on soda is related to the number of cans you purchase.

   a. Show the relationship as a table of values containing 5 sets of values.

   b. Make a graph based on the table of values.

   c. Write an equation to express the relationship between the number of cans and the cost.

Solution

a. Table of values

Number of Cans	Cost
1	$0.75
2	$1.50
3	$2.25
4	$3.00
5	$3.75

b. Graph

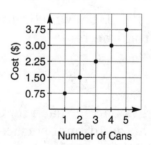

c. Equation

To obtain the cost (C), multiply the number of cans by the cost of one can.

$C = 0.75n$, where n is the number of cans.

2. Prior to the holiday season, a department store advertises a special offer:

> For every $50 spent, receive a $10 gift card for future purchases.

Make a graph to represent the relationship described by this offer.

Solution Think about the relationship between how much you spend and how many gift cards you receive. For a purchase under $50, you do not receive a gift card. Notice that you receive one gift card if you spend any amount between $50 and $99.99. For purchases between $100 and $149.99, you receive two gift cards, and so on.

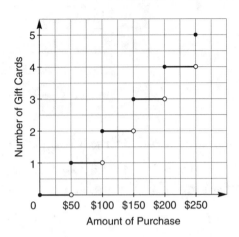

Special Offer

Note: This graph represents a **step function**.

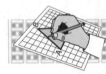 **PRACTICE**

1. This table indicates a linear relationship between x and y.

x	1	3	5	7	9
y	1	7	13	?	25

According to this pattern, which number is missing from the table?

A. 15   B. 19
C. 21   D. 23

2. A plumber charges $48 for each hour she works plus an additional service charge of $25. At this rate, how much would the plumber charge for a job that took 4.5 hours?

3. The cost of a long-distance telephone call can be computed based on the formula $T = C + nr$, where

T = total cost of the call in dollars
C = charge for the first three minutes in dollars

n = number of additional minutes the call lasts
r = rate per minute for each additional minute in dollars

What is the cost of a 15-minute long-distance call if a person is charged $1.75 for the first three minutes and $0.15 for each additional minute?

4. A local parking lot charges $1.75 for the first hour and $1.25 for each additional hour or part of an hour. Represent the relationship of parking charges to hours parked:

a. in a table of values for 1 to 6 hours.
b. in an equation in which t represents time parked in hours and C represents the total cost of parking.
c. as a graph with hours parked on the horizontal axis and total cost on the vertical axis.

Assessment Macro A

1. What is the next term in this sequence?

$$2, 3\frac{1}{4}, 4\frac{1}{2}, 5\frac{3}{4}, \ldots$$

A. $6\frac{1}{4}$ B. $6\frac{1}{2}$ C. $6\frac{3}{4}$ D. 7

2. What is the 30th term of this arithmetic sequence?

$$4, 9, 14, 19, 24, \ldots$$

A. 124   B. 129   C. 149   D. 154

3. Which of these sequences is a geometric sequence?

A. 2, 4, 6, 8, 10,…
B. 2, 4, 8, 16, 32,…
C. 2, 4, 8, 32, 256,…
D. 2, 4, 6, 10, 16,…

4. Which of the following terms could NOT be a term of the sequence $\frac{1}{4}, \frac{1}{2}, 1, 2, \ldots$?

A. 16   B. 32   C. 84   D. 128

5. Which of the following numerical patterns is equivalent to this visual pattern?

A. 0.33333…   B. 123123…
C. 133133133…   D. 313131…

6. Suppose this pattern were continued:

A	BA	BBA	BBBA
	AB	BAB	BBAB
		ABB	BABB
			ABBB

How many B's will be in the 20th diagram?

A. 20   B. 60   C. 360   D. 380

7. What is the units digit in 3^{47}?

A. 1   B. 3   C. 7   D. 9

8. What is the pattern of the units digits in the sequence $8^1, 8^2, 8^3, 8^4, \ldots, 8^n$?

A. 8, 4, 6, 2   B. 8, 6, 4, 2
C. 8, 2, 4, 6   D. 8, 4, 2, 6

9. In January, a New Jersey Devils poster sells for $10.00. In February, the price increases 10%. In March, it decreases 10%. In April, it increases 10%, and so on. (It continues to alternate between the 10% increase and 10% decrease.) Which of the following graphs could represent the situation described?

A.

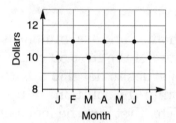

B.

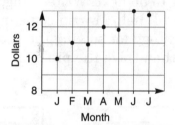

C.

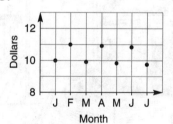

D.

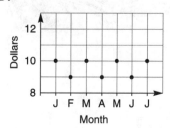

10. Examine the pattern:

    MONDAYMONDAYMONDAY…

If this pattern is continued, what letter will be in the 121st position?

  A. M   B. N   C. D   D. Y

11. In an arithmetic sequence, the 9th term is 27 and the 11th term is 33. Find the common difference.

12. 3, 9, 27,… are the first three terms of a geometric sequence. What is the 8th term?

13. If the visual pattern is continued, how many dots will be needed to represent the ninth term?

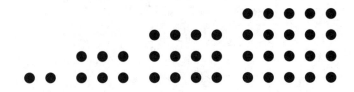

14. What is the next term of the sequence 5, 7, 12, 19, 31,…?

15. What digit is in the 52nd decimal place in the decimal value for $\frac{3}{11}$?

Open-Ended Questions

16. Martin's homework assignment is to find the sum of the first 50 odd numbers. He decides to develop a pattern to help obtain the solution to this problem. The first sum he looks at is $1 + 3 + 5 + 7 = 16$. Martin is not sure what to do next. Write a suggestion that will help Martin come up with a strategy for solving the problem, and then solve the problem yourself.

17. Suppose the pattern shown were continued.

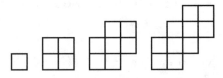

  a. Extend the pattern for two more terms.
  b. Discuss how the pattern is produced.
  c. How many squares would be in the 100th diagram?

18. A rectangle with an area of 24 square units can have dimensions 8×3. Consider other possible sets of dimensions for rectangles with an area of 24 square units.

  a. Draw a graph of length vs. width for all rectangles with sides that are whole numbers whose areas equal 24 square units. (Put length on the x-axis and width on the y-axis.)
  b. Based on the graph, what happens when the length gets extremely large?

Use algebraic concepts and processes to concisely express, analyze, and model real-world situations.

4 B 1 Expressions and Open Sentences

A **variable** is a letter used to represent a number. The value of a variable can change. A **term** is a number, a variable, or the product or quotient of a number and a variable. A **variable expression** is made up of one or more terms.

Variable Expressions	Constant Terms	Numerical Expressions
$6a$	7	7×3
$x + 4$	4π	$\dfrac{4 + 5}{3}$
a^2b	2	$6(2 + 5 - 3)$

The value of an expression involving a variable depends upon the value used for the variable.

When evaluating an expression be sure to follow the established algebraic **order of operations**:

- Perform any operation(s) inside parentheses, brackets or above or below the division bar.
- Simplify any terms with exponents.
- Multiply and divide in order from left to right.
- Add and subtract in order from left to right.

Examples

a.
$$9 - 3 \times 2 + 8$$
$$= 9 - 6 + 8$$
$$= 3 + 8$$
$$= 11$$

b.
$$12^2 - 9 \times 5 + \frac{18}{10-7}$$
$$= 12^2 - 9 \times 5 + \frac{18}{3}$$
$$= 144 - 9 \times 5 + \frac{18}{3}$$
$$= 144 - 45 + 6$$
$$= 105$$

Model Problem

1. If ⬜ represents x^2, ⬛ represents x, and □ represents 1, write the expression represented by:

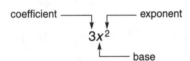

Solution Since ⬜ = x^2, ⬛ = x, and □ = 1, the expression is $3x^2 + 2x + 4$.

2. Evaluate $3a^2 + 4ab - b^2$ for $a = 2$ and $b = -1$.

Solution Substitute the values for a and b into the expression.

$$3(2)^2 + 4(2)(-1) - (-1)^2$$
$$= 3 \times 4 + (-8) - 1$$
$$= 12 - 8 - 1$$
$$= 4 - 1$$
$$= 3$$

In a term that contains a variable, the numerical factor is called the **coefficient**.

$$\text{coefficient} \longrightarrow \quad \longleftarrow \text{exponent}$$
$$3x^2$$
$$\longleftarrow \text{base}$$

If terms have exactly the same variables raised to the same powers, they are called **like terms**.

Like Terms	Not Like Terms
$2x$ and $7x$	$2x$ and $2y$
$5x^2$ and $\frac{1}{2}x^2$	$3x^2$ and $3x^3$
$5a^2b$ and $6a^2b$	$5ab$ and $6ac$

If an expression contains like terms, the terms can be combined to simplify the expression.

To combine like terms:

- Add (or subtract) the coefficients.
- Carry along the like base and exponent.

Model Problem

Simplify $5x^2 + 8y + 2x^2 - 5y$.

Solution Combine like terms.

$$5x^2 + 8y + 2x^2 - 5y$$
(with $3y$ combining $8y$ and $-5y$, and $7x^2$ combining $5x^2$ and $2x^2$)

$$= 5x^2 + 2x^2 + 8y - 5y$$
$$= 7x^2 + 3y$$

Verbal statements can be written as variable expressions.

To translate from a verbal statement to a variable expression:
- Identify the variable(s) to be used.
- Using the key words, determine the operation(s) involved.
- Form the expression, using symbols of grouping as needed.

Model Problem

Write a variable expression for each of the verbal statements:

a. the product of two consecutive integers
b. 10 less than twice a number

Solution

a. $n(n + 1)$
b. $2n - 10$

Variable expressions can be used to translate word problems into equations or inequalities, which are also called **open sentences.**

To translate a word problem into an open sentence:
- Write a variable expression to represent each word phrase.
- Determine the appropriate relationship symbol to be used ($=, \neq, <, >, \leq, \geq$).

Word Phrase	Symbol
is equal to	=
is not equal to	≠
is greater than	>
is less than	<
is at least	≥
is greater than or equal to	≥
is at most	≤
is less than or equal to	≤

Model Problem

1. Translate each verbal sentence into an equation or inequality.

a. What number doubled gives a result of 18?

b. Three times one more than a number gives a result of at least 26.

c. What number squared gives a result of 81?

d. Four times three more than a number gives a result of 100.

e. What number tripled gives a result that is less than 27?

Solution

a. $2x = 18$
b. $3(n + 1) \geq 26$
c. $x^2 = 81$
d. $4(n + 3) = 100$
e. $3r < 27$

2. Translate the verbal expression into an inequality.

> One more than five times a number gives a result between 4 and 16 (including 4 and 16).

Solution The fact that the result is between 4 and 16 means that it is greater than or equal to 4, but less than or equal to 16.

One more than 5 times a number is greater than or equal to 4 translates to

$$5n + 1 \geq 4.$$

One more than 5 times a number is less than or equal to 16 translates to

$$5n + 1 \leq 16.$$

The two inequalities combined can be written as

$$4 \leq 5n + 1 \leq 16.$$

PRACTICE

1. Evaluate the expression $a^2 b^3$ when $a = 2$ and $b = -1$.

A. 4   B. −4   C. −32   D. −64

2. Evaluate $5 + x(x + 2)$ when $x = 8$.

A. 71   B. 85   C. 122   D. 130

3. If $n + 7$ is an even number, the next larger even number is

A. $n + 5$ B. $n + 9$
C. $10n + 7$ D. $2n + 14$

4. Which of the following cannot be simplified?

A. $2a^2 + 5a^2$ B. $3a^2 - 3a$
C. $5a - 11a$ D. $16a^3 - 6a^3$

5. Which expression must be added to $2x - 4$ to produce a sum of 0?

A. 0 B. $x + 2$
C. $2x + 4$ D. $-2x + 4$

6. The perimeter of the parallelogram is $6a + 8b$. Find the length of each of the other two sides.

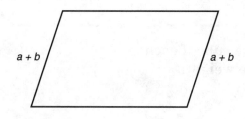

A. $2a + 3b$ B. $2b + 3a$
C. $4a + 6b$ D. $3a + 4b$

7. Write an open sentence to represent the following statement.

If 7 is subtracted from 4 times a certain number, the difference is 25.

A. $7 - 4n = 25$ B. $4(n - 7) = 25$
C. $4(7 - n) = 25$ D. $4n - 7 = 25$

8. Write an open sentence to represent the following statement.

5 less than twice a number is 3 more than the number.

A. $2n - 5 = 3n$
B. $2n - 5 = n + 3$
C. $5 - 2n = n + 3$
D. $5 - 2n = 3n$

9. Write an equation to show that segment y is 5 units longer than 3 times the length of segment x.

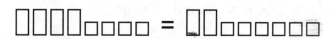

A. $y = 15x$ B. $y = 3x + 5$
C. $x = 3y + 5$ D. $y = 3x - 5$

10. Find the value of A in the formula $A = \frac{1}{2}bh$, if $b = h = 5$.

11. A rectangle has dimensions $2x$ by $x + 3$. Write an expression for the perimeter of the rectangle. Combine any like terms.

12. Write the open sentence represented by the following situation. The congruent sides of an isosceles triangle have lengths that are each 5 inches greater than the length of the base. The perimeter is at most 31 inches.

13. Is the expression for twice the sum of a number and 10 the same as the expression for the sum of twice a number and 10? Explain.

14. Write the equation modeled by the following diagram.

$$\square\square\square\square\square\square\square\square = \square\square\square\square\square\square\square\square$$

4 B 2 Linear Equations and Inequalities

An equation is a statement that two expressions are equal to each other.

An equation is similar to a balanced scale.

To determine the unknown weight in a balanced scale:

- Remove (cancel) identical items from both sides of the scale.
- Determine a relationship among the remaining items.
- Substitute the value for the known quantity.
- Determine the weight of the unknown quantity.

 Model Problem

Given the balances as shown, find the weight of one cube if each ball weighs 1 pound and the cubes are all the same weight.

1.

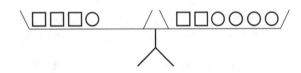

Solution

- Remove two cubes and one ball from each side of the scale.
- One cube is balanced by three balls.
- Each ball weighs 1 pound.
- One cube weighs 3 pounds.

2.

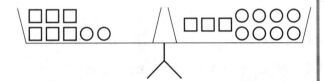

Solution

- Remove three cubes and two balls from each side of the scale.
- Three cubes are balanced by six balls.
- Each ball weighs 1 pound.
- Three cubes weigh 6 pounds, and one cube weighs 2 pounds.

The more traditional method of solving an equation involves use of mathematical properties.

To solve an equation:

- Remove parentheses by multiplication. (Apply the distributive property. See page 15.)
- Combine like terms on each side of the equation. Use addition or subtraction as indicated.
- Collect the variable terms on one side, and the number terms on the other side. Use the opposite operation of the one indicated (the *inverse* operation) to move a term from one side of the equation to the other.
- Rewrite the variable term with a coefficient of 1. Use the opposite operation of the division or multiplication indicated.

Model Problem

1. Solve for x: $6x = 2(x + 1) + 10$.

Solution

$$6x = 2(x + 1) + 10$$
$6x = 2x + 2 + 10$ Remove parentheses.
$6x = 2x + 12$ Combine like terms.
$6x - 2x = 2x - 2x + 12$ Collect the variable terms on one side.
$4x = 12$ Combine like terms.
$\dfrac{4x}{4} = \dfrac{12}{4}$ Rewrite the variable term with a coefficient of 1.
$x = 3$ Divide.

2. Solve for y: $5y + 7 = 2(y + 5)$.

Solution

$$5y + 7 = 2(y + 5)$$
$5y + 7 = 2y + 10$ Remove parentheses (using distributive property).
$5y - 2y + 7 = 2y - 2y + 10$ Collect the variable terms on one side.
$3y + 7 = 10$
$3y + 7 - 7 = 10 - 7$ Collect the number terms on the other side.
$3y = 3$
$\dfrac{3y}{3} = \dfrac{3}{3}$ Rewrite the variable term with a coefficient of 1.
$y = 1$ Divide.

Inequalities

An **inequality** consists of two or more terms or expressions connected by an inequality sign. The solution to an inequality is given as a solution set and can be represented on a number line.

The process of solving an inequality is very similar to the process of solving an equation. The major difference is that if you are multiplying or dividing the inequality by a *negative number*, you must reverse the order of the inequality sign.

Note: On the number line, a closed circle indicates the value is included in the solution set. An open circle indicates the value is excluded from the solution set.

Model Problem

1. Graph the inequality on the number line: $-2 < x \le 4$.

Solution It may help to read one side of the inequality at a time.

The expression $-2 < x$ means "x is greater than -2."
The expression $x \le 4$ means "x is less than or equal to 4."

The combined meaning is "x is greater than -2 and less than or equal to 4." Therefore, x is between -2 and 4, with -2 excluded and 4 included. Since -2 is excluded, put an open point on -2. Since 4 is included, put a closed point on 4. All points in between -2 and 4 are included. On the number line the graph would be:

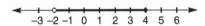

2. Solve the inequality for x and graph the solution on a number line.

$$-4x + 4 \le -16$$

Solution

$$-4x + 4 \le -16$$
$$-4x + 4 - 4 \le -16 - 4 \qquad \text{Add } -4.$$
$$-4x \le -20$$

$$\frac{-4x}{-4} \ge \frac{-20}{-4} \qquad \text{Divide by } -4; \text{ reverse inequality sign.}$$

$$x \ge 5$$

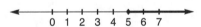

3. Write an inequality to describe the following situation and solve it, then graph the inequality.

"Seven less than twice a number is greater than –3."

Solution

$$2x - 7 > -3$$
$$2x > 4$$
$$x > 2$$

1. Within the set of integers, which of the following represents the solution of the inequality?

$$3x + 1 < 8$$

A. $\{..., -3, -2, -1, 0, 1, 2\}$
B. $\{..., -3, -2, -1, 0, 1, 2, 3\}$
C. $\{..., -3, -2, -1, 0, 1, 2, 3, 4, 5, 6, 7\}$
D. $\left\{..., -3, -2, -1, 0, 1, 2, 2\frac{1}{3}\right\}$

2. Write an equation that expresses the following statement:
"Five more than three times a number equals twice the number."

A. $3(x + 5) = 2x$ B. $3x + 5 = x^2$
C. $3x + 5 = 2x$ D. $3(x + 5) = x$

3. Which equation has NO integer as a solution?

A. $3x = 9$ B. $16x = 32$
C. $3x = 2$ D. $17x = 51$

4. Solve for c: $\dfrac{c}{6} + 14 = 38$.

A. 4   B. 144
C. 214   D. 312

5. Which number line shows the graph of $-3 \leq x < 4$?

A.

B.

C.

D.

6. Solve for x: $-5x + 3 \geq 28$.

7. Write an inequality to describe the following situation and solve:
Nine more than half a number is at most -8.

8. Solve for x: $2(x + 1) - 4 = x + 3$.

9. How do you know that the solution to the equation $3x = 251$ is NOT an integer?

10. Given the balance shown, find the weight of each identical cube if each ball weighs 1 pound.

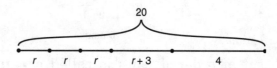

11. Solve for r.

12. The temperature at High Point is 28°C and is dropping at the rate of 1.5° per hour.
The temperature at Belmare is about 18°C and is rising at the rate of 2° per hour.

a. Write an expression representing the temperature at each place after x hours.
b. Write an equation to represent that both cities are at the same temperature.
c. Solve the equation to find out how many hours (to the nearest tenth) it will take for the two cities to be at the same temperature.

4 B 3 Functions

Recall that the formula for the circumference of a circle is $C = \pi d$. The circumference *depends* on the length of the diameter. Another way to express the relationship is to say, "Circumference is a **function** of diameter." The value of the function for different values of d, can be displayed by a table, a set of ordered pairs, an equation, and a graph.

a. Table:

d	1	2	3	4
C	3.14	6.28	9.42	12.56

(d and C in inches)

b. Set of ordered pairs:

$$\{(1, 3.14), (2, 6.28), (3, 9.42), (4, 12.56),\ldots\}$$

c. Equation:

$$C = \pi d$$

d. Graph:

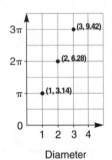

Diameter

A set of ordered pairs is called a **relation**. The previous example shows that circumference and diameter constitute a relation. You can observe that for any specified value for the diameter, there can be only one resulting value for the circumference. A relation with this property is called a **function**.

When a relation is listed as a set of ordered pairs, a function exists when for every x-value (first coordinate) there is only one y-value (second coordinate).

Model Problem

Which of the following is NOT a function?

A. $(-1, 5), (1, 4), (5, 3), (0, 6)$ B. $(2, 4), (4, 8), (9, 18)$

C. $(3, 0), (3, 1), (3, 2), (3, 3)$ D. $(9, 0), (7, 0), (-5, 0), (13, 0)$

Solution   To be a function a given set of ordered pairs must be such that for every value of x there is only one value for y. In choice C, 3 is repeated for x with four different y-values. Hence, C is not a function.

Example

Consider the following relationship: "*y* equals 4 more than *x*."

Equation: $y = x + 4$

Table of values:

x	*y*
−2	2
−1	3
0	4
1	5
2	6

Graph:

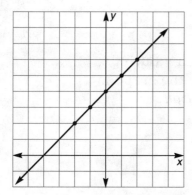

In this example, *y* is clearly a function of *x*. For any given value of *x*, the value of *y* is found by adding 4 to the value of *x*. This is an example of a linear function. Additional examples of linear functions include:

$y = 5$ $y = -x$ $x + y = 5$

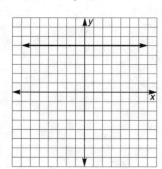 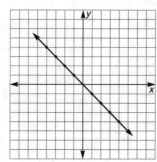 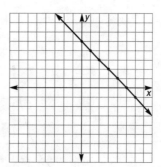

Example

Consider the following relationship: "y equals three more than the square of x."

Equation: $y = x^2 + 3$

Table of values: Graph:

x	y
0	3
1	4
−1	4
2	7
−2	7

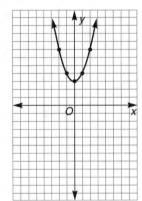

Set of ordered pairs:
{(0, 3), (1, 4), (−1, 4),
(2, 7), (−2, 7),...}

In this example, y is clearly a function of x. For any given value of x, the value of y is found by adding 3 to the square of x. Since the graph is not a straight line, this is an example of a *non-linear function*.

Example

An additional example of a non-linear function would be:

Equation: $y = 2^x$.

Table of values: Graph:

x	y
0	1
1	2
2	4
3	8
4	16

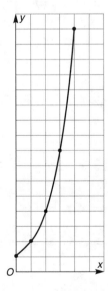

In this example, y is a function of x. For any given value of x, y is found by raising 2 to the power of x. The shape of the graph confirms that the function is non-linear.

Model Problem

The garage at the movie theater charges $3.00 for the first hour (or any part of the hour) of parking and $1.00 for every additional hour (or part of the hour).

a. How much would the charge be for 4 hours and 20 minutes?
b. Dave and Jay park their car in the garage at 2:00 P.M. At 3:55 P.M., when their movie is finished, Dave wants to run back to pick up the car so that they can save money. Jay says a few minutes don't matter, so they should take their time. Who is correct?
c. Graph the function from part b.

Solution

a. 4 hours and 20 minutes would be the same charge as 5 hours. With $3.00 for the first hour and 4 hours at $1.00 per hour, the final charge would be $7.00.
b. Dave is correct. The charge for the first hour is $3.00. The charge for the second hour is $1.00. The second hour ends at 4:00 P.M. If they pick up the car after 4:00 P.M., they will be charged for three hours.

c.

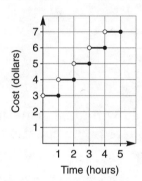

PRACTICE

1. Which of the following sets of ordered pairs does NOT represent a function?

   A. {(−2, 2), (2, 2), (−5, 5), (5, 5)}
   B. {(7, 5), (8, 5), (−8, 5), (−7, 5)}
   C. {(2, 3), (2, 4), (2, 5), (2, 6)}
   D. {(1, 1), (2, 8), (3, 27), (4, 64)}

2. Which is NOT the graph of a function?

A.
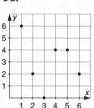

B.
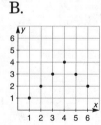

C.
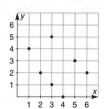

D.
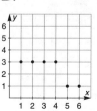

3. A function is described by the equation $R = 2t + 3$. Complete the table by finding the value of R when $t = 10$.

t	1	3	3	...	10
R	5	7	9	...	?

A. 11   B. 15
C. 23   D. 26

4. Which of the following equations represents the line containing the points given in the graph?

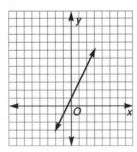

A. $y = x + 1$ B. $y = x + 2$
C. $y = 2(x + 1)$ D. $y = 2x + 1$

5. Which of the following tables would be a reasonable representation of the relationship between the price (P) of an item and the sales tax (T) for that item?

A.

P	T
2.00	0.12
6.00	0.12
15.50	0.12
85.00	0.12

B.

P	T
2.00	0.48
6.00	0.36
15.50	0.24
85.00	0.12

C.

P	T
2.00	0.12
6.00	0.36
15.50	0.93
85.00	5.10

D.

P	T
2.00	0.12
6.00	0.14
15.50	0.24
85.00	0.12

6. This graph illustrates a constant function:

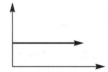

Which of the following is a situation that can be modeled by a constant function?

A. the number of cars at a parkway tollbooth during the day
B. the ratio of the circumference to the diameter of every tire in an auto shop
C. a car's speed during a 20-minute drive to the store
D. the value of a car over a 10-year period

7. The given table is generated from which of the following rules?

x	-2	-1	0	1	2
y	-2	-3	-4	-5	-6

A. $y = 2x + 2$ B. $y = -4 - x$
C. $y = x + 4$ D. $y = -3x$

8. Callie purchases a laptop computer on an installment plan. She pays $50 a month until the computer is paid off. Which of the following graphs matches the relationship between months and the unpaid balance?

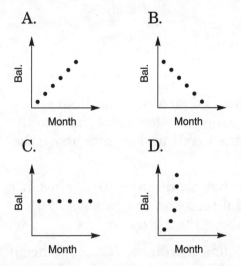

A.
B.
C.
D.

9. This graph illustrates a step function:

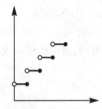

Which of the following applications could NOT be represented by the graph?

A. postage rates per ounce
B. charges at a parking lot per hour
C. taxicab fare compared to distance
D. temperature over the course of an afternoon.

10. Karen and Samantha were looking at the table showing selected New Jersey towns and cities with their zip codes.

Town or City	Zip Code
Little Falls	07424
West Paterson	07424
Wayne	07470
West Orange	07052
Newark	07101
Newark	07102
Newark	07103

The two girls agreed that the relationship of town or city to zip code was not an example of a function. Karen said the Little Falls/West Paterson example was the reason. However, Samantha said that it was the Newark example that resulted in it not being a function. Who is correct? Explain your position.

4 B 4 Cartesian Coordinate System

The **Cartesian coordinate system** (also known as the **coordinate plane**) is formed by two perpendicular number lines. The horizontal number line is called the **x-axis** and the vertical number line is called the **y-axis**. The two axes separate the plane into four **quadrants**. The point where the two axes intersect is called the **origin**, and this point is represented by the ordered pair (0, 0).

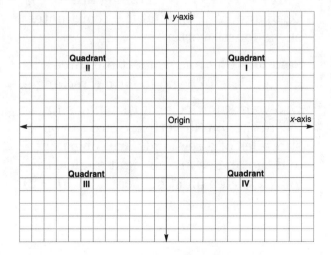

Given an **ordered pair**, (x, y), any point can be graphed in the coordinate plane. The first number in the ordered pair is the **x-coordinate**. The second number in the ordered pair is the **y-coordinate**. It is important to know that point $(3, 7)$, for example, is in a different location than point $(7, 3)$.

 # Model Problem

Give the coordinates of the points shown:

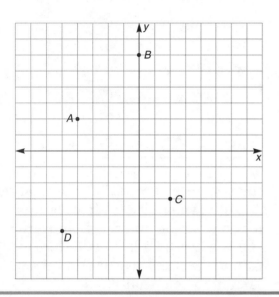

Solution Point A has coordinates $(-4, 2)$ since it is located 4 units to the left of the y-axis and 2 units above the x-axis.

Point B has coordinates $(0, 6)$ since it is located directly on the y-axis and 6 units above the x-axis.

Point C has coordinates $(2, -3)$ since it is located 2 units to the right of the y-axis and 3 units below the x-axis.

Point D has coordinates $(-5, -5)$ since it is located 5 units to the left of the y-axis and 5 units below the x-axis.

The rectangular coordinate system can be used to represent geometric situations, algebraic solutions and displays of data.

 # Model Problem

Three vertices of a rectangle are $(0, 6)$, $(2, 6)$ and $(2, -2)$. Find the coordinates of the fourth vertex.

Solution Plot and connect the three given points to begin to form a rectangle.

Use your knowledge of rectangles to locate the fourth vertex. Since the opposite sides of a rectangle are congruent and the angles are all right angles, the fourth vertex must be located at $(0, -2)$.

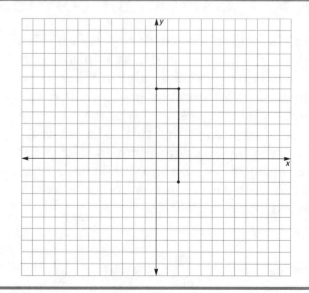

1. Segment AB is horizontal. What must be the coordinates of point B if the length of $\overline{AB}$ is 10 units?

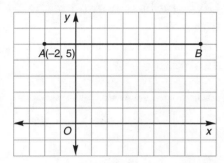

   A. (10, 0)   B. (10, 5)
   C. (8, 5)   D. (12, 5)

2. If you plot the following points on a grid, what kind of triangle is formed? $A(0, 0)$, $B(8, 0)$, $C(10, 5)$

   A. right   B. acute
   C. obtuse   D. isosceles

3. A rectangle located in the first quadrant has two vertices at (0, 0) and (0, 6). Where will the other two vertices be if the area of the rectangle is 30 square units?

   A. (0, 5) and (6, 5)
   B. (5, 0) and (6, 5)
   C. (5, 0) and (5, 6)
   D. (9, 0) and (9, 6)

4. What is the best name for quadrilateral $ABCD$ defined by $A(0, 0)$, $B(8, 0)$, $C(10, 5)$, and $D(2, 5)$?

   A. rectangle
   B. trapezoid
   C. isosceles trapezoid
   D. parallelogram

5. Which of the following points is not in the second quadrant?

   A. (−7, 1)   B. (−80, 80)
   C. (−5, −5)   D. (−5, 1.5)

6. A square has vertices at (−2, 2), (2, 2), (2, −2), (−2, −2). Which of the following is NOT true?

   A. The point (1, 1) is inside the square.
   B. The point (2, −3) is inside the square.
   C. The point (−2.5, 0) is outside the square.
   D. The point (1.5, 2) is on the square.

7. Which of the ordered pairs satisfies all of the clues?
   Clue 1: The point is closer to the x-axis than the y-axis.
   Clue 2: The absolute value of the x-coordinate is 6.
   Clue 3: Each coordinate is a negative number.

   A. (−6, −1)   B. (−6, 1)
   C. (1, −6)   D. (−1, −6)

8.

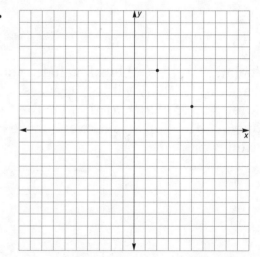

   If a line is drawn through the points (2, 5) and (5, 2), the line will pass through which quadrants?

   A. Quadrants I, II, and III
   B. Quadrants I, II, and IV
   C. Quadrants I and III
   D. Quadrants I, III, and IV

9. a. Find the values of y in the equation $y = 5x + 1$, for the x-values $-1, 0,$ and 2.

b. Express the results from part a in the form of ordered pairs.

c. Plot the points on a graph. What do you observe about the points?

10. Given $A(1, 5)$, $B(1, -1)$, $C(9, -1)$.

a. Plot the points on a grid, and form a figure.

b. Give the best name for the figure formed.

c. Find the area enclosed by the figure.

4 B 5 Rates of Change

Consider the linear function $y = 2x + 3$.

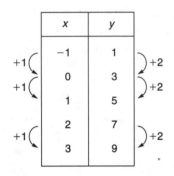
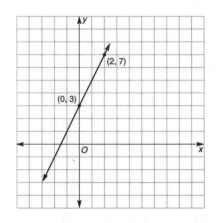

By looking at the table of values, you can see that as the x-value increases by 1, the y-value increases by 2. This means that there is a constant rate of change (ratio) for the function.

Contrast the linear function above with the following non-linear function, $y = x^2$.

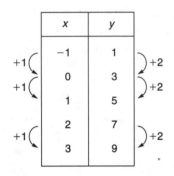

By looking at the table of values, you can see that as the x-value increases by 1, the y-value decreases and then increases by different amounts. This means that the function does not have a constant rate of change.

Model Problem

The table shows the relationship between the length of a side of a square and the perimeter.

Side Length	1	2	3	4	5
Perimeter	4	8	12	16	20

Graph the data and determine the rate of change between the length of a side and the perimeter.

Solution   Record side length on the horizontal axis and perimeter on the vertical axis. Graph the data as ordered pairs.

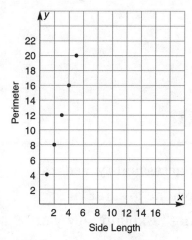

Choose any two points from the data.

(1, 4)   side length = 1   perimeter = 4
(2, 8)   side length = 2   perimeter = 8

$$\text{ratio} = \frac{\text{difference in perimeter}}{\text{difference in side length}}$$

$$\text{ratio} = \frac{8 - 4}{2 - 1} = \frac{4}{1} = 4$$

Answer   The rate of change is 4. This will be true no matter which two points you use.

Slope

Slope is a number indicating the steepness of a line. A basic property of a straight line is that it has a constant slope. This constant slope expresses the constant rate of change of a linear function.

Since line l is steeper than line m, the slope of line l is a larger number than the slope of line m.

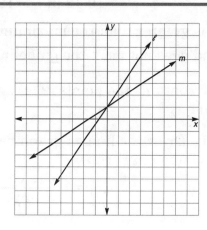

If the coordinates of any two points on a line are known, the slope can be found by finding the following ratio:

$$\text{slope} = \frac{\text{rise}}{\text{run}} = \frac{\text{vertical change}}{\text{horizontal change}} = \frac{\text{change in } y}{\text{change in } x}$$

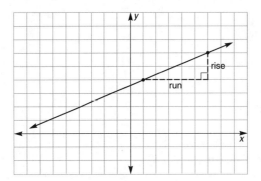

Model Problem

1. Many buildings have ramps for wheelchair accessibility. Federal guidelines require that the ramp have a slope no greater than $\frac{1}{12}$. Does the ramp shown meet the guidelines? Explain how you determined your answer.

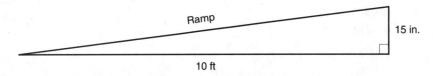

Solution The slope of the ramp is given by the ratio,

$$\frac{\text{rise}}{\text{run}} = \frac{15 \text{ in.}}{10 \text{ ft}}$$

$$= \frac{15 \text{ in.}}{10(12 \text{ in.})} = \frac{15 \text{ in.}}{120 \text{ in.}} = \frac{1}{8}$$

Since $\frac{1}{8}$ is greater than the maximum allowed, $\frac{1}{12}$, the ramp is too steep and does not meet the guidelines.

2. A line contains the points (2, 4) and (7, 7). What is the slope of the line?

Solution Draw the line on graph paper. Find the change in y (rise) and the change in x (run).

$$\text{slope} = \frac{\text{change in } y}{\text{change in } x} = \frac{3}{5}$$

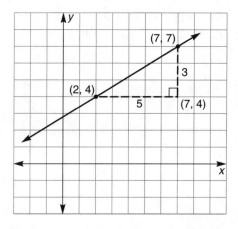

Note: In finding the change in y and the change in x, it is helpful to draw in the small right triangle as illustrated above.

Lines with Different Types of Slopes

If a line is parallel to the x-axis, the line has no steepness. Its slope is zero.
If a line is parallel to the y-axis, the line has an undefined slope.
If a line rises from left to right, its slope is positive.
If a line falls from left to right, its slope is negative.

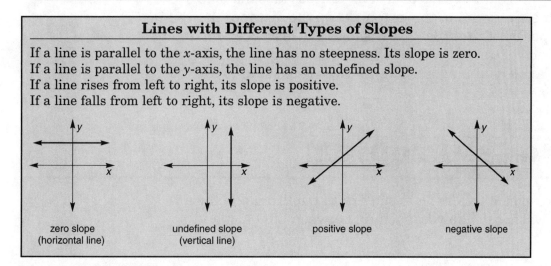

Picturing Relationships

Graphs can be used to illustrate relationships and to show the way quantities change over time. The following graph shows the relationship between the distance covered by a car driving in the city and the amount of gas remaining in the gas tank. (Assume that the car is always moving.)

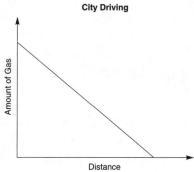

The horizontal axis represents distance traveled and the vertical axis represents the amount of gas remaining in the gas tank. As the car travels a greater distance, the amount of gas remaining decreases.

The graphs of certain real-world situations may contain different sections. Each section communicates a different portion of the relationship.

Carlos's Ride

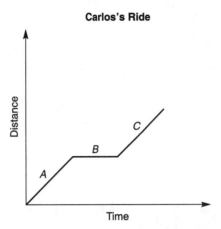

This graph depicts the distance traveled as Carlos rides his bike over some period of time. Section *A* of the graph shows the distance increasing over the first period of time. Section *B* shows the distance remained the same for a period of time. (Carlos stopped for some reason.) Finally, section *C* shows Carlos resuming his bike ride at a rate equal to the rate from section *A*.

 Model Problem

The graph corresponds to Mrs. Johnson's auto trip from one town to another. What was most likely happening between 2:00 P.M. and 2:30 P.M.?

Mrs. Johnson's Drive

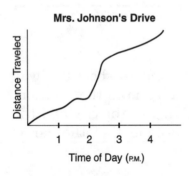

A. Mrs. Johnson was in heavy traffic.
B. Mrs. Johnson stopped for lunch.
C. Mrs. Johnson was looking for a parking space downtown.
D. Mrs. Johnson was driving on the highway.

Solution   Notice that there is a large increase in the distance traveled from 2:00 P.M. to 2:30 P.M., as compared to the distance traveled from 1:00 P.M. to 2:00 P.M. The only choice that would match this large increase in distance is D. Since it is possible to drive at higher speeds on the highway, Mrs. Johnson was able to travel a large distance during the specified half hour.

PRACTICE

1. A line with a positive slope passes through the point $(0, 4)$. Which of the following points could be on the line?

   A. $(3, 3)$ B. $(2, 0)$

   C. $(10, 10)$ D. $(-2, 5)$

2. A line passing through the point $(3, 1)$ has a slope of zero. Which of the following points could be on the line?

   A. $(-3, 1)$ B. $(-3, 0)$

   C. $(3, 0)$ D. $(3, 6)$

3. What is the slope of the line shown?

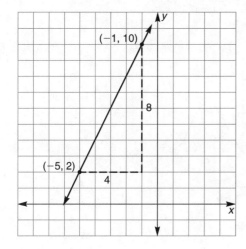

   A. -2 B. $-\dfrac{1}{2}$ C. $\dfrac{1}{2}$ D. 2

4. How many lines in the figure have negative slope?

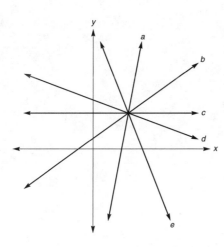

   A. 0   B. 1   C. 2   D. 3

5. The following graphs represent Andy riding his bike. Which graph shows Andy accelerating, traveling at a constant speed, and then quickly stopping?

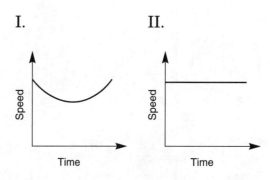

   A. I   B. II   C. III   D. IV

6. The graph shows the speed of a cyclist on an afternoon bike ride. What is happening at B?

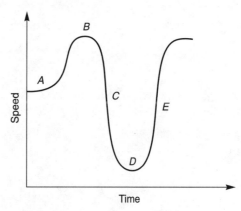

   A. She is riding up a hill.
   B. She is changing direction.
   C. She is reaching her greatest speed.
   D. She is stopping for a rest.

7. Alyssa bicycled for 2 hours at 4 miles per hour. She stopped for one hour to visit a friend. She bicycled for another hour at 5 miles per hour. Which graph best represents Alyssa's trip if the horizontal axis is time and the vertical axis is distance?

A.

B.

C.

D.

8. Which graph best depicts population increasing over a period of years.

A.

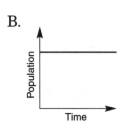

B.

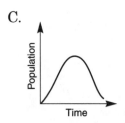

C.

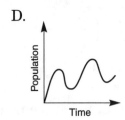

D.

Assessment Macro B

1. What is the weight of one cube if each pyramid weighs 3 pounds and the cubes are all the same weight?

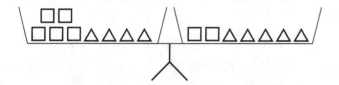

  A. $\frac{1}{3}$ pound   B. 1 pound
  C. 3 pounds   D. 9 pounds

2. Evaluate the expression $\frac{a}{b} + b$ for $a = 16$ and $b = -2$.

  A. 6   B. -6
  C. -10 D. -16

3. The inequality $-3x > 12$ is equivalent to which of the following?

  A. $x < -4$ B. $x > -4$
  C. $x < 4$ D. $x > 4$

4. Find the solution to this inequality.

$$3(x - 1) - 7 < x - 3$$

  A. $x < 0.5$ B. $x < 2.5$
  C. $x < 3.5$ D. $x < 6.5$

5. Which of these expressions have been combined correctly?

  I. $3x^2 + 5x + 9x^2 - 4x = 12x^2 + 1$
 II. $5ab + 3c^2 - 2ab + 8c^2$
     $= 3ab + 11c^2$
III. $8p^3 + 8p^2 + 8p + 8$
     $= 8p^3 + 8p^2 + 8p + 8$

  A. I and II only   B. I and III only
  C. II and III only   D. I, II, and III

6. Which of the following is NOT a function?

  A. $\{(2, 4), (3, 4), (4, 4), (5, 4)\}$
  B. $\{(0, 1), (3, 4), (6, 7), (9, 10)\}$
  C. $\{(-5, 0), (-4, 4), (4, 10), (5, 11)\}$
  D. $\{(-2, 4), (-2, 3), (-2, -3), (-2, -4)\}$

7. A price p is increased 10%. Which of the following is NOT a representation for the new price?

  A. $p + 0.1p$ B. $110\% \cdot p$
  C. $1.1p$ D. $0.1p$

8. The area of a trapezoid $= \frac{1}{2}h(b_1 + b_2)$. Which of the following expressions represents the area of the trapezoid shown?

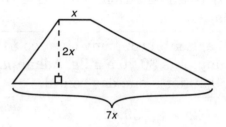

  A. $6x^2$ B. $8x^2$ C. $10x^2$ D. $16x^2$

9. Within the set of integers, which of the following represents the solution of the inequality?

$$4x + 1 < -9$$

  A. $\{-1, 0, 1, 2, 3,...\}$
  B. $\{-2, -1, 0, 1, 2, 3,...\}$
  C. $\{..., -5, -4, -3\}$
  D. $\{..., -5, -4, -3, -2\}$

10. If you plot the following points on a grid, what kind of triangle is formed?

$$A(0, 0), B(-4, 0), C(-1, -10)$$

  A. right   B. acute
  C. obtuse   D. isosceles

11. Which of the following sentences does the equation $5n - 3 = n + 11$ correctly represent?

  A. 3 less than 5 times a number is 11.
  B. 5 times 3 less than a number is 11 more than the number.
  C. 3 more than 5 times a number is 11 more than the number.
  D. 3 less than 5 times a number is 11 more than the number.

12. Which of the following represents the situation described?

> Joyce had $950 in her savings account. She withdrew the same amount each month for 5 months. After depositing $100 in her account, the balance was $908. How much money did she withdraw each month?

    A. $950 - 5x = 908 + 100$
    B. $950 + 5x - 100 = 908$
    C. $950 - 5x + 100 = 908$
    D. $950 - 5x - 100 = 908$

13. Write an equation to describe the situation.

> A car rental company charges $15 a day plus $0.20 a mile. Adam paid $59.40 for a one-day rental.

    A. $0.20x = \$59.40$
    B. $0.20x - 15 = \$59.40$
    C. $0.20x + 15 = \$59.40$
    D. $0.20x + 15 + \$59.40$

14. Which inequality matches the situation?

> Take a number, add 3, multiply by 3, and subtract twice the original number. The result is greater than 5.

    A. $3x + 3 > 5 - 2x$
    B. $3x + 3 - 2x > 5$
    C. $3(x + 3) - 2x > 5$
    D. $9x - 2x > 5$

15. Which number line shows the graph of $-4 < x \leq 4$?

    A.

    B.

    C.

    D.

16. A function is described by the equation $V = t^2 + 2t$. What value of V is missing from the table?

t	-2	-1	0	1	2	3	...	10
V	0	-1	0	3	8	15	...	?

    A. 40   B. 100
    C. 120   D. 200

17. Which of these three equations have the same solution?

    I. $4x = 32$

    II. $\dfrac{1}{4}x = \dfrac{1}{32}$

    III. $40x = 320$

    A. I and II only   B. I and III only
    C. II and III only   D. I, II, and III

18. A vertical line segment has one endpoint at $(1, -3)$ and the other endpoint in the fourth quadrant. What are the coordinates of the other endpoint if the length of the segment is 10 units?

    A. $(1, 10)$ B. $(1, 7)$
    C. $(1, -10)$ D. $(1, -13)$

19. A parking garage charges $2.00 for the first hour and $1.25 for each additional hour of parking. The parking fees are given by the formula:

$$F = 2.00 + 1.25(h - 1),$$

where h is the number of hours and F is the total fee. What value completes the table showing parking fees?

h	1	2	4	...	10
F	2.00	3.25	5.75	...	?

20. Solve for a.

Open-Ended Questions

21. The graph shows the cost of electricity:

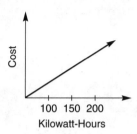

Cost

100 150 200
Kilowatt-Hours

a. Which of the following statements is NOT true concerning this graph?

A. The graph illustrates a function.
B. The graph shows that the cost of electricity levels off at some point.
C. As the number of kilowatt-hours increases, the cost also increases.
D. The cost increases at the same rate between 100 and 150 kilowatt-hours and between 150 and 200 kilowatt-hours.

b. Select one of the statements that is true. Explain why it is true.
c. For the statement that is false, explain why it is false.

22. The table shows a linear relationship between x and y.

x	y
1	3
2	5
3	7
4	9

Based on the indicated relationship:

a. Provide three additional pairs of values in the table.
b. Graph the relationship on a coordinate grid.
c. Express the relationship between x and y as an equation.

23. Read the following number puzzle.

> Start with a number.
> Multiply it by 4.
> Add 6.
> Divide by 2.
> Subtract twice your original number.

a. Express each step of the puzzle as an algebraic expression.
b. Explain why the answer to the puzzle will always be 3.

24. The graph shows the cost of an item over a long period of time.

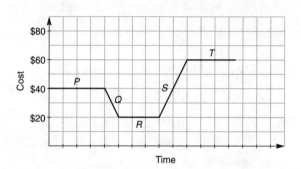

Cost

$80
$60
$40
$20

P
Q
R
S
T

Time

a. Describe section P of the graph. What does it tell you about the cost of the item?
b. Name another section of the graph with the same type of result as section P. How is this section the same as section P? How is this section different from section P?
c. Describe section Q of the graph. What does it tell you about the cost of the item over a more limited period of time?

1. Which of the following is NOT a geometric sequence?

   A. 1, 1, 1, 1, 1,...
   B. 10, 100, 1,000, 10,000,...
   C. 6, 4, 2, 0, −2,...
   D. 6, 3, 1.5, 0.75,...

2. Which of the following cannot be simplified?

   A. $11a^2 - 13a^2$ B. $11a^2 - 10a$
   C. $4a^2 + 11a^2$ D. $4a + 11a$

3. What digit is in the 20th decimal place in the decimal value of $\dfrac{35}{101}$?

   A. 3   B. 4   C. 5   D. 6

4. What is the units digit in 12^{15}?

   A. 2   B. 4   C. 6   D. 8

5. Given the pattern TEXASTEXAS-TEXAS..., find the letter in the 99th position.

   A. A   B. T   C. X   D. S

6. Which graph represents $x - y = 5$?

   A.
   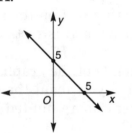

   B.
   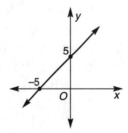

   C.
   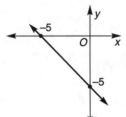

   D.
   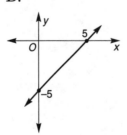

7. The table represents the function $y = -2x - 2$. What value do you get for y when $x = -6$?

x	y
0	−2
1	−4
2	−6

   A. 14   B. 10
   C. −10   D. −14

8. Solve for x: $\dfrac{x}{2} + \dfrac{3x}{2} = 10$.

   A. $x = 2.5$ B. $x = 5$
   C. $x = 10$ D. $x = 20$

9. A sequence is generated by the rule $3n^2 - 4$, where n represents the number of the term in the sequence. What is the difference in the values of the 25th and 26th terms in the sequence?

   A. 6   B. 153
   C. 159   D. 459

10. Which of the following is a translation for "6 less than 3 times a number"?

    A. $6 < 3n$ B. $3n > 6$
    C. $3n - 6$ D. $3(n - 6)$

11. Write and simplify an expression for the surface area of this rectangular prism.

    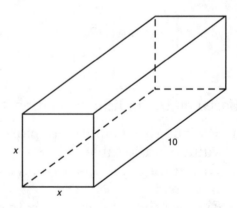

    A. $2x^2 + 10x$ B. $2x^2 + 40x$
    C. $40x$ D. $44x$

12. In which of the inequalities would you need to reverse the inequality symbol when solving?

A. $-5x > -35$ B. $6y \le 20$

C. $x + 16 \ge -14$ D. $\dfrac{1}{5}x \le 6x$

13. If the horizontal axis represents time and the vertical axis represents price, which graph shows the sharpest increase in price over time?

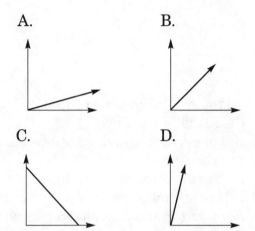

A.

B.

C.

D.

14. What is the difference between the 15th terms of Sequence A and Sequence B?

Sequence A: 2, 4, 8, 16, …
Sequence B: −2, 4, −8, 16, …

A. 0   B. 16,384
C. 32,768   D. 65,536

15. The number of mold cells on a piece of bread doubles every 12 minutes. If there are 35 mold cells on the bread now, about how many cells will there be 2 hours from now?

A. 420   B. 840
C. 2,458   D. 35,840

16. Suppose you start with $39.65 in your bank. Each day you put in $1.35 more than you put in on the previous day. That is, on day 1 you put in $1.35, on day 2 you put in $1.35 + $1.35 or $2.70, on day 3 you put in $4.05, and

so on. How much money will you have in the bank on the 12th day?

A. $55.85   B. $128.75
C. $144.95   D. $492

17. At Global Packages, shipping charges are $4.25 for the first 3 pounds and 75¢ for each additional pound. At that rate, how much did a package weigh if the charges were $11?

A. 6 pounds   B. 9 pounds
C. 12 pounds   D. 15 pounds

18. Which of the following lines has a negative slope?

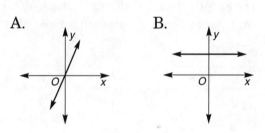

A.

B.

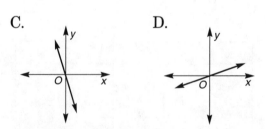

C.

D.

19. Which of the following points is NOT on the graph of $3x + y = 15$?

A. (5, 0)   B. (−5, 0)
C. (3, 6)   D. (6, −3)

20. Which of the following equations represents the line containing the points given in the graph?

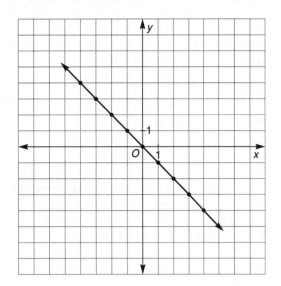

A. $y = x$
B. $y + x = 0$
C. $x - y = 0$
D. $y = x - 1$

21. Each table gives the coordinates of three points on different lines. Which one represents a line with zero slope?

A.

x	y
2	-3
2	0
2	3

B.

x	y
-2	2
0	4
2	6

C.

x	y
-2	7
0	7
2	7

D.

x	y
-2	-7
0	-5
2	-3

22. A hill rises 3 meters vertically for every 30 meters of horizontal distance. What is the slope of the hill?

A. $\dfrac{1}{10}$
B. $\dfrac{10}{3}$
C. 10
D. 27

23. Solve for x: $4(x + 2) - 2(x - 3) = 20$.

A. $x = 10.5$
B. $x = 9$
C. $x = 7.5$
D. $x = 3$

24. Evaluate $3 - 5c^3$ when $c = 2$.

A. -27
B. -37
C. -120
D. -997

25. The graph shows a relationship between distance and time. Which statement is FALSE?

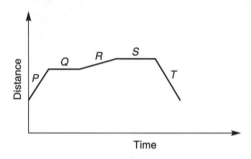

A. Segment T shows a gradual decrease in distance.
B. Segment Q shows distance remaining constant.
C. Segment R shows a gradual increase in distance.
D. Segment P shows a sharp increase in distance.

26. From the given clues, identify the endpoints of a line segment.

Clue 1: One endpoint is in the second quadrant.
Clue 2: The other endpoint is in the third quadrant.
Clue 3: The length of the line segment is 10 units.
Clue 4: Each endpoint is 4 units from the y-axis.
Clue 5: The endpoint in the third quadrant is 3 units from the x-axis.

27. Evaluate the expression $a^b + b^a$ when $a = 5$ and $b = 2$.

28. The figure is made of nine congruent squares. If the total area is 144 square units, solve for x.

29. Solve for r.

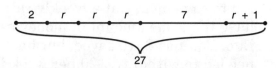

30. A display of Frisbees has been hung on a wall in the shape of a triangle. There is 1 Frisbee in the top row, 2 Frisbees in the second row, 3 in the third, and so on, with each row containing one more Frisbee than the row above. The display contains 12 rows. How many Frisbees are used in the entire display?

31. Consider the following pattern:

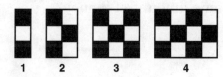

If the pattern is continued, how many small squares would be shaded in the 9th diagram?

32. A special sequence is formed by taking 5 more than the sum of the two previous terms to find the third term and all succeeding terms. If the first four terms of the sequence are 1, 2, 8, 15, … , find the 10th term.

33. The first four terms of an arithmetic sequence are 3, 7, 11, 15, and 123 is the 31st term. What is the value of the 30th term?

Open-Ended Questions

34. a. Plot the ordered pairs (8, 0), (7, 1), (6, 2), and (5, 3).
   b. Sketch the graph suggested by the ordered pairs.
   c. Describe the pattern in words.
   d. What equation describes the pattern?

35. Describe a situation that would match each graph shown. Indicate what would be measured on each axis.

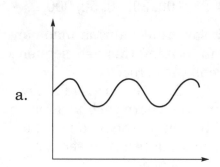

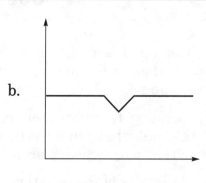

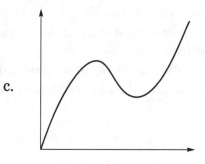

36. Consider the pattern shown.

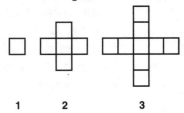

1   2   3

   a. If the pattern were extended, how many small squares would there be in the 20th picture?

   b. Explain why the number of small squares would always be 1 more than a multiple of 4.

37. A balanced scale is shown. The boxes are of equal weight. Each ball weighs 1 kg. Find the weight of one box. Explain how you got your answer.

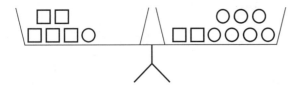

38. Jake and Ben are twins. Jake is trying to save money at a weekly rate to have the same amount of money as Ben. Ben has $310 saved, but has needed to withdraw $20 per week to help meet his expenses. Jake has $100 to start and adds $10 per week to the amount.

   a. Write an expression that represents the amount of money Jake will have after x weeks.

   b. Write an expression that represents the amount of money Ben will have after x weeks.

   c. At this rate, how many weeks will it take until Jake and Ben have the same amount of money?

Extra Practice

Open-Ended Questions

Study Questions 1 and 2, which are presented with their solutions, and then try Questions 3 and 4.

When responding to an open-ended question, think about what you must do to form a response that will receive a score of 3.

- Answer all parts of the question.
- Present your work clearly, so that the person grading it will understand your thinking.
- Show all your work, including calculations, diagrams, and written explanations.

1. Given two sequences as follows:

     A: 5, 10, 15, 20, 25, 30,…

     B: 10, 20, 30, 40, 50, 60,…

- Sabrina knows that the 70th term of sequence B is 700. How can she quickly obtain the value of the 70th term of sequence A? Explain your approach.
- Sabrina forms a new sequence C by squaring each term of sequence A. She comes up with:

     C: 25, 100, 30, 40, 50, 900,…

Derek says that Sabrina made an error in forming this new sequence. Where is the error?

- Show the first six terms of a new sequence D where each term is found by taking the mean of the corresponding terms in sequence A and sequence B.

Solution (for a score of 3):

Sabrina can quickly find the 70th term of sequence A by observing that each term in sequence A is one-half of the corresponding term in sequence B. The 70th term of sequence A would be 350 since the 70th term of sequence B is 700.

Sabrina squared three of the terms correctly: 5, 10, and 30. Her errors were that she doubled 15, 20, and 25 instead of squaring them. The correct sequence C is

C: 25, 100, 225, 400, 625,...

It may be helpful to use a table to organize your calculations for sequence D.

Term	A	B	Mean of A and B	D
1	5	10	15 ÷ 2	7.5
2	10	20	30 ÷ 2	15
3	15	30	45 ÷ 2	22.5
4	20	40	60 ÷ 2	30
5	25	50	75 ÷ 2	37.5
6	30	60	90 ÷ 2	45

D: 7.5, 15, 22.5, 30, 37.5, 45

2. The rectangle shown has a perimeter given by the expression $6x + 2$.

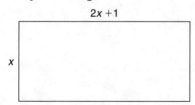

- If the perimeter equals 32 inches, what is the area of the rectangle?
- A square has a perimeter given by the expression $100x + 60$. Write an expression for the length of one of the sides of the square.
- Give two sets of expressions for the length and width of a rectangle if the perimeter of the rectangle is given by the expression $20x + 8$.

Solution (for a score of 3)

To calculate the area, you must first find the length and width. Solve the equation for x.

$$\text{Perimeter} = 6x + 2 = 32$$
$$6x = 30$$
$$x = 5$$

So, the length and width are

$$l = 2(5) + 1 = 11$$
$$w = 5$$
$$\text{Area} = lw = 5(11) = 55 \text{ square units}$$

Since the perimeter of a square is equal to $4s$, dividing the perimeter by 4 will result in an expression for a side of the square.

$$4s = 100x + 60$$
$$s = \frac{100x + 60}{4}$$
$$s = 25x + 15$$

The perimeter of a rectangle is expressed by the formula

$$2(l + w).$$

Setting this equal to the expression given, we get

$$2(l + w) = 20x + 8.$$

Simplify:

$$(l + w) = 10x + 4.$$

So, any two expressions for length and width that have a sum of $10x + 4$ will result in perimeter of $20x + 8$.

Possible answers are:

Expressions	Check
$l = 8x + 3$ $w = 2x + 1$	$(8x + 3) + (2x + 1) = 10x + 4$ $2(10x + 4) = 20x + 8$
$l = 7x + 2$ $w = 3x + 2$	$(7x + 2) + (3x + 2) = 10x + 4$ $2(10x + 4) = 20x + 8$
$l = 9x + 4$ $w = x$	$(9x + 4) + x = 10x + 4$ $2(10x + 4) = 20x + 8$

3. Given the two sequences as follows:

A: 1, 2, 4, 8, 16,...

B: 0, 50, 100, 150, 200,...

- Which sequence will exceed 1,000 first? Show how you found your answer.
- Find the number of the term where sequence A becomes greater than sequence B.
- Sequence C starts at 24 and each term is found by multiplying the preceding term by $\frac{1}{2}$. Show the first five terms of sequence C. At what term number does sequence C become less than 1?

4. Rectangle $ABCD$ has dimensions $l = 4x$ and $w = x$. The perimeter of $ABCD$ is given by the expression $10x$.

- If the perimeter of $ABCD$ equals 60 centimeters, what is its area?
- Square $LMNO$ has a perimeter given by the expression $16x + 4$. Write an expression for the length of one side of the square.
- A rectangle and a square both have a perimeter of 20 cm, but they have different areas. Draw a possible square and a possible rectangle with their dimensions labeled.

PART 1   MULTIPLE-CHOICE QUESTIONS

1. The number of Central High School students who gave to the United Fund this year was 342. This figure is 110% of what it was the previous year. This means that

   A. 10 more Central High School students gave to the United Fund.
   B. The number of Central High School students giving to the United Fund decreased from last year to this year.
   C. Central High School raised more money for the United Fund this year than it did last year.
   D. The number of Central High School students giving to the United Fund increased from last year to this year.

2. Which point on the number line below could represent the product of the numbers represented by W and X?

   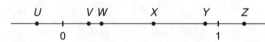

   A. U B. V C. Y D. Z

3. Which figure does NOT have symmetry?

   A.   B.

   C.   D.

4. Which of the following points lies on the graph of $3x - y = 6$?

   A. $(-1, -3)$ B. $(0, 6)$
   C. $(1, -3)$ D. $(1, 3)$

5. A circle has a center at the point $(4, 4)$ and a radius of 3 units. Which of the following points is outside the circle?

   A. $(4, 6)$ B. $(6, 4)$
   C. $(0, 0)$ D. $(1, 4)$

6.

   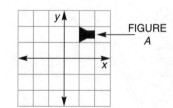

   Which of the following represents the result of reflecting figure A in the y-axis and then reflecting that image in the x-axis?

   A.   B.

   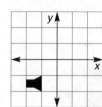   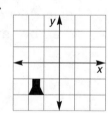

   C.   D.

   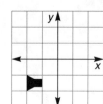   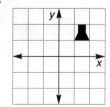

7. Start with the set of whole numbers between 10 and 40, including 10 and 40.

10 11 12 13 14 15 16 17 18 19
20 21 22 23 24 25 26 27 28 29
30 31 32 33 34 35 36 37 38 39 40

Remove all prime numbers.
Remove all perfect squares.
Remove all factors of 72.
Remove all multiples of 9.
Remove all numbers in the Fibonacci sequence 1, 1, 2, 3, 5, 8, . . .

How many numbers remain?

A. 11　　B. 12　　C. 13　　D. 14

8. The baselines of a baseball diamond form a square with side lengths of 90 feet.

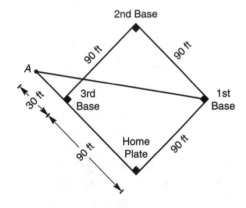

Zach catches a ball at point A, which is on the foul line 30 feet beyond 3rd base. How far from 1st base is Zach when he catches the ball?

A. 120 ft　　　B. 150 ft
C. 180 ft　　　D. 210 ft

9. The table below indicates a relationship between a and b.

a	0	1	2	•	•	5	•	•	8
b	−2	1	4	•	•	13	•	•	22

Which of the equations below expresses the relationship between a and b that is indicated in the table?

A. $b = a^2$　　　　B. $b = 2a + 3$
C. $b = 3a - 2$　　D. $b = a + 3$

10. Each month, Orchard Middle School does a feature newspaper article on one of its students, who is picked at random. The numbers of male and female students in each grade at Orchard Middle School are shown in the table below.

Number of Students							
Grade 5		Grade 6		Grade 7		Grade 8	
M	F	M	F	M	F	M	F
28	22	31	20	25	26	25	23

Based on this table, what is the probability that the student chosen will be a female student in grade 5?

A. .11　　B. .22　　C. .24　　D. .44

11. For a sale, a shopkeeper lowers the original price of an item by 20 percent. After the sale, the shopkeeper raises the price of that item by 20 percent of its sale price. The price of the item is then

A. more than the original price
B. less than the original price
C. the same as the original price
D. There is not enough information to compare the two prices.

12. How many one-eighths would equal the sum of 48 one-halves and 36 one-fourths?

A. 30　　　　　B. 84
C. 264　　　　D. 1,920

13. Which of the following graphs most likely shows the relationship between a used car's resale value and its age?

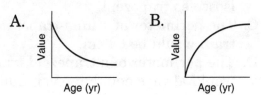

A.

B.

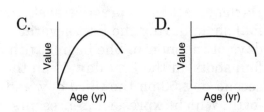

C.

D.

14. Which of the following sets of lengths does NOT represent a triangle?

A. {1, 1, 1}   B. {6, 8, 10}
C. {2, 3, 5}   D. {9, 12, 15}

15. Which of the following has a value that is less than zero?

A. 2.04×10^2
B. 2.04×10^0
C. 2.04×10^{-3}
D. All three of the above are greater than or equal to zero.

16. In the spinner shown, if the probability of obtaining A on one spin is $\frac{1}{12}$, how many degrees are contained in the central angle for A?

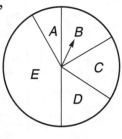

A. 20°   B. 30°   C. 45°   D. 60°

17. Which of these statements is TRUE?

A. All squares are similar.
B. All triangles are similar.
C. All rectangles are similar.
D. All parallelograms are similar.

18. If the pattern shown is continued, what letter will be in the 99th position?

TRIANGLETRIANGLETRIANGLE ...

A. E   B. T   C. I   D. N

19. 1,376 students attending Cary Junior High School all voted for president of the student council. With approximately one-fifth of the votes counted, the leading candidate had 185 votes. Assuming that candidate obtained the same proportion of the total number of votes, the total number of votes he or she received would be between

A. 250 and 300   B. 450 and 500
C. 600 and 700   D. 900 and 950

20. Based on the graph shown, which car takes the least amount of time to go from 0 to 60 mph, and about how much less time does it take?

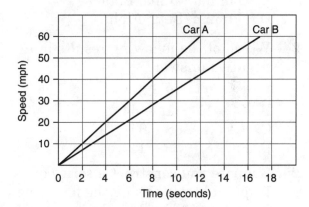

A. car A by about 5 seconds
B. car B by about 5 seconds
C. car A by about $2\frac{1}{2}$ seconds
D. car B by about $2\frac{1}{2}$ seconds

21. The cost of first-class postage changed from 34¢ to 37¢. This change translates into what percent increase? (Round your answer to the nearest tenth.)

A. 3%   B. 3.7%   C. 8.1%   D. 8.8%

22. The whole numbers from 1 to 36 are each written on a small slip of paper and placed in a box. If one slip of paper is selected at random from the box, what is the probability that the number selected is a factor of 36 and also a multiple of 8?

A. $\frac{13}{36}$ B. $\frac{1}{36}$ C. $\frac{1}{4}$ D. 0

23. Sal's Drugstore is having a sale of photographic film.

Film Sale	
a roll of ALPHA film	20-exposure film for $2.30
a roll of BETA film	12-exposure film for $1.50
a roll of GAMMA film	30-exposure film for $3.15

Place the brands of film in order from least cost per exposure to greatest cost per exposure.

A. Alpha, Beta, Gamma
B. Beta, Alpha, Gamma
C. Gamma, Alpha, Beta
D. Gamma, Beta, Alpha

24. Which of the following results in a negative answer?

A. $12 \div 2 - 18 \div 3$
B. $44 \div 11 - 11 \div 44$
C. $8 \div 88 - 88 \div 8$
D. $88 \div 8 - 88 \div 11$

25. Suppose you construct a series of trapezoid trains using the following isosceles trapezoid block:

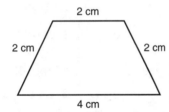

A two-trapezoid train looks like:

A three-trapezoid train looks like:

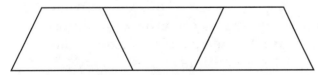

Which of the following statements is NOT going to be true?

A. An eight-trapezoid train would be a parallelogram.
B. A nine-trapezoid train would be an isosceles trapezoid.
C. The perimeter of a nine-trapezoid train would be 64 cm.
D. The perimeter of a trapezoid train would always equal an even number of centimeters.

26. Darlene decides to practice basketball foul shots during the first several days of the month. She begins with 30 foul shots on the first day, 40 on the second day, 50 on the third day, and so forth. Which expression represents the number of practice foul shots on the ninth day?

A. $30 + 9(10)$ B. $30 + 8(10)$
C. $30(10)^9$ D. $30 + 10(10)$

27. While interviewing students at Washington Middle School, Paul asked 20 students (picked at random) what their favorite fall sport is. Maria asked 50 different students (picked at random) the same question. George combined Paul's data and Maria's data. All three graphed their results.

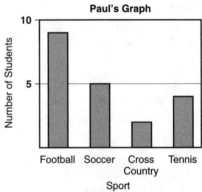

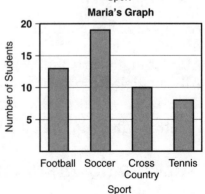

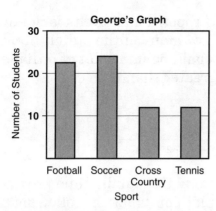

George's Graph

Which of these graphs could be used to give the most reliable estimate of the percentage of the Washington Middle School student population whose favorite fall sport is soccer?

A. Paul's
B. Maria's
C. George's
D. There is no reason to use one graph rather than another.

28. At what point does the graph of $3x - 2y = 6$ cross the x-axis?

A. $(0, -3)$　　　B. $(-3, 0)$
C. $(0, 2)$　　　D. $(2, 0)$

29. Which of these pieces of cardboard cannot be folded along the dotted lines to make a closed box?

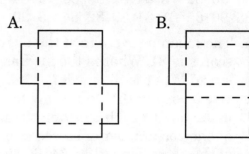

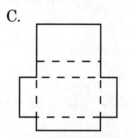

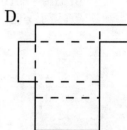

30. What is the weight of one of the cubes if each pyramid weighs 2 pounds?

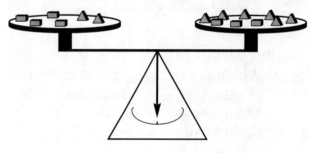

A. $\frac{1}{4}$ pound　　　B. 4 pounds
C. 6 pounds　　　D. 8 pounds

PART 2 OPEN-ENDED QUESTIONS

DIRECTIONS: Respond in detail to each open-ended question. Show your process and provide an explanation as directed in the question. Your score for each question will be based on the accuracy and completeness of your process or explanation.

31. Use a centimeter ruler to determine the lengths of the sides of the figure below. Sketch the figure and label the corresponding sides of your sketch with the measurements you find. Use these measurements to find the perimeter and the area of the figure below. Show all work clearly.

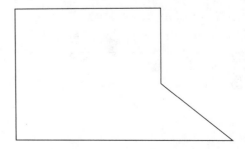

32. The math exam scores for the 21 students in Mr. Walker's homeroom were:

65 90 82 78 84 92 88 86 70 68 75
88 90 85 61 81 79 82 84 83 90

a. The mean or average of the above scores is 81. What is the median score? What is the mode?

b. Use a copy of the grid below to make a bar graph showing the frequency or number of scores in each of the score ranges 60–64, 65–69, 70–74, etc.

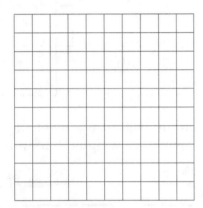

c. Which is the best general indicator of this class's performance on the exam: the mean, median, or mode? Explain your answer.

33. Your task is to form 3-digit numbers from the six digits in the box. Repetition of digits is not allowed.

$$3 \quad 4 \quad 5 \quad 6 \quad 7 \quad 8$$

a. How many 3-digit numbers can be formed?

b. How many 3-digit numbers would exist if you were allowed to repeat digits? Explain your process.

c. If repetition of digits is not allowed, how many of the 3-digit numbers would be multiples of 5 that are greater than 500?

34.

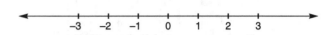

a. Copy the number line provided and plot points for the following numbers. Label each point.

$$\frac{3}{4}, 0.5$$

b. Name two different numbers that are greater than 0.5 and less than $\frac{3}{4}$. Write one of your numbers in fractional form and write the other number in decimal form.

c. Explain how you know that each of your numbers is greater than 0.5 and less than $\frac{3}{4}$.

35. Two shipping companies charge different amounts to mail packages. Company A charges an initial $5 fee, and each pound shipped is an additional $1. Company B charges an initial $3 fee, and $1.50 for each pound shipped.

a. How much would each company charge to mail a package weighing 2 pounds?

b. For what weight package will the two companies charge the same amount?

c. Which company charges less for a 6 lb package? How much will you save by choosing this company to ship your 6 lb package?

PART 1 MULTIPLE-CHOICE QUESTIONS

1. Three students started at the same flagpole in the middle of a large, flat, grassy area and chose three different directions in which to walk. Each walked for 10 yards in a straight line away from the pole. Suppose many more students did this, each walking in a direction different from the directions chosen by all the others. If you think of the final positions of the students as points, which of the following figures would contain all of those points?

   A. triangle   B. square
   C. rhombus   D. circle

2. If $3y - 25 = 3 - 11y$, then $y =$

   A. -3.5 B. -2
   C. 2 D. 3.5

3. On Mondays and Tuesdays, the social studies teacher Ms. Hardaway has the students in class 812 work in groups of four. On Wednesdays and Thursdays, they work in groups of six. On Fridays, they work in groups of eight. On any day that all the students are present, there is always one student left over after the groups are formed. Which of the following could be the number of students in class 812?

   A. 37   B. 33   C. 29   D. 25

4. All numbers that have 2, 5, and 11 as factors will also have which number as a factor?

   A. 18   B. 25   C. 55   D. 220

5. Performing which set of transformations on the white figure below will NOT result in the white figure covering the black figure completely?

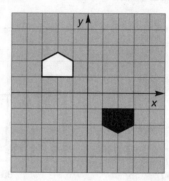

   A. reflection in the y-axis followed by translation 4 units down
   B. reflection in the y-axis followed by reflection in the x-axis
   C. translation 4 units to the right followed by reflection in the x-axis
   D. rotation of 180° about the origin

6. Curley Middle School offers three extracurricular activity groups: marching band, glee club, and science club. There are some students in each of these groups. Every student belongs to at least one of them, but some belong to more than one. Which of the following diagrams would best represent this situation?

   A.    B.

   C.   D.

7. Four friends are planning to eat at a restaurant where complete dinners cost between $12.00 and $17.00 per person. They want to leave the waiter a tip amounting to 15% of their total bill. Which of the following is the closest to what the four friends will need to leave for their combined TOTAL tip?

A. $2.00   B. $4.00
C. $9.00   D. $15.00

8. Fifteen students' scores on their last math test are represented in the bar graph shown.

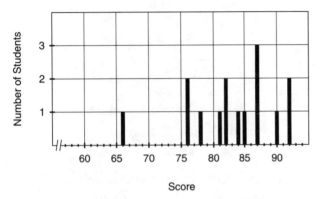

What are the mean, median, and mode for the set of scores represented in the graph?

A. mean = 83, median = 84,
 mode = 87
B. mean = 84, median = 83,
 mode = 87
C. mean = 83, median = 87,
 mode = 84
D. mean = 84, median = 87,
 mode = 83

9. Consider the number 36. Some pairs of matching positive integer factors of 36 are (2, 18), (3, 12), (4, 9), (6, 6), (12, 3), and (18, 2). Suppose a student graphed all possible pairs of matching positive integer factors of a given positive integer. Which of the following most likely represents such a graph?

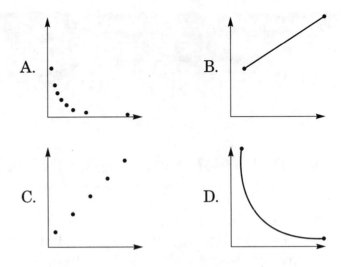

A.   B.

C.   D.

10. Which point on the number line shown could represent the product of U and X?

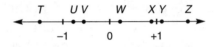

A. point T B. point V
C. point W D. point Y

11. A diagram of a rectangle with dimensions 6 inches by 8 inches is placed in a copy machine, which is set to enlarge all dimensions by 10 percent. Will the resulting figure fit on an $8\frac{1}{2}$" × 11" sheet of paper?

A. Yes, it will fit with room to spare.
B. Yes, it will just fit with no room to spare.
C. No, one dimension will fit, but not the other.
D. No, both dimensions will be too large.

12. For 45 cents, a snack-food vending machine dispenses a small bag of Crunchy Corn, which weighs one and one-eighth ounces. At this rate, the

cost of one pound of Crunchy Corn would be between

A. $2.30 and $2.75
B. $4.30 and $4.75
C. $6.30 and $6.75
D. $8.30 and $8.75

13. The graph shows a relationship between distance and time. Which statement is FALSE?

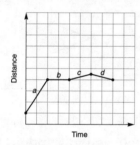

A. Segment a shows a sharp increase in distance.
B. Segment b shows distance remaining constant.
C. Segment c shows a gradual increase in distance.
D. Segment d shows a sharp decrease in distance.

14. The length $\overline{AD}$ in the diagram is between which two integers?

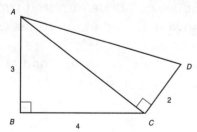

A. 3 and 4   B. 4 and 5
C. 5 and 6   D. 6 and 7

15. The rectangular prism shown has 6 faces. Which of the following would NOT be an expression for the area of one of the faces?

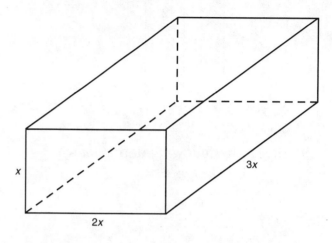

A. $6x^2$ B. $4x^2$
C. $3x^2$ D. $2x^2$

16. Triangle MNP is an isosceles triangle with a vertex angle of 40°. If $\overline{AB} \parallel \overline{DC} \parallel \overline{NP}$, what is the measure of $\angle ABC$?

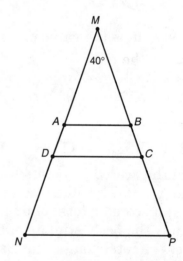

A. 40°   B. 70°
C. 110°   D. 140°

17. The tenth term of the geometric sequence $\frac{1}{3}$, 1, 3, 9, 27, . . . is

A. 2,187   B. 6,561
C. 19,683   D. 59,049

18. Which of the following equations gives the rule for the relationship in the table?

x	y
0	−2
1	1
2	4
3	7

A. $y = x - 2$ B. $y = 2x$
C. $y = 3x - 2$ D. $y = x + 4$

19. Given the following definitions:

 means $x + 3$

 means x^2

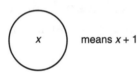 means $x + 1$

What would be the value of

 + ?

A. 13 B. 15 C. 23 D. 26

20. Samantha scored 16 points in a basketball game. The other members of the team scored a total of 62 points. Which is the best estimate of the percent of the total points that Samantha scored?

A. 15% B. 20% C. 25% D. 30%

21. Suppose the U.S. Postal Service has proposed increasing the cost of first-class postage to 40¢ for the first ounce or fraction thereof and 28¢ for each additional ounce or fraction thereof. Which of the graphs best represents the cost of mailing a first-class item depending on the weight of that item in ounces?

A.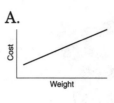

B.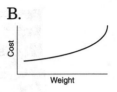

C.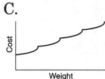

D.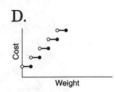

22. Which of the following would NOT be a net for a pyramid?

A.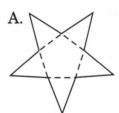

B.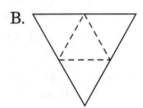

C.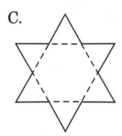

D.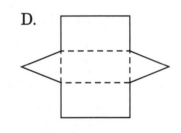

23. A number is chosen at random from a set containing the first 30 even counting numbers: {2, 4, 6, . . . , 56, 58, 60}. What is the probability that the number has 11 as the sum of its digits?

A. $\frac{1}{30}$ B. $\frac{1}{15}$

C. $\frac{1}{10}$ D. $\frac{2}{5}$

24. Each of the students in a class rolled a number cube one hundred times and graphed how many times each number came up. (Each face of the cube is labeled with just one of the digits 1, 2, 3, 4, 5, 6.)

Which of the following graphs most likely represents the one the students made of the results of their whole class?

A.

B.
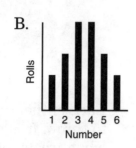

C.
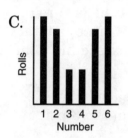

D.
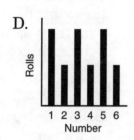

25. You are playing a game in which you move a chip on a number line. Where you move the chip is determined by the cards you draw from a pack. Each card has an integer printed on it. Your chip is now at the position shown below.

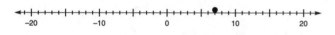

On each turn, you move your chip to the location with the coordinate equal to the sum of the coordinate of your current location and the number on the card you draw. Suppose you draw cards with the following sequence of numbers: −2, 6, −7, −12, and −4. At what coordinate is your chip located after you complete this sequence of moves?

A. 12　　　　B. 11
C. −12　　　D. −19

26. The Burger Baron Restaurant is open from 6 A.M. until midnight and serves all meals. Every half hour during an 8-hour period last Tuesday, Ronald

counted the number of customers in the restaurant. He graphed his data, but forgot to label the time-of-day axis.

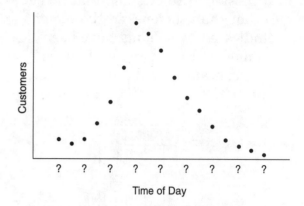

Which of the following time-of-day axes is most likely labeled the way it should have been in Ronald's graph?

A.	3 P.M.	4	5	6	7	8	9	10	11
B.	10 A.M.	11	12 P.M.	1	2	3	4	5	6
C.	12 P.M.	1	2	3	4	5	6	7	8
D.	6 A.M.	7	8	9	10	11	12 P.M.	1	2

27. Keisha works for a florist making bouquets. On a given day the florist has daisies, carnations, roses, lilies, and tulips in stock. How many combinations of flowers can Keisha have for bouquets if she wants to include at least three different types of flowers?

A. 10　　B. 15　　C. 16　　D. 60

28. A card is drawn at random from a standard deck of 52 playing cards. The card is put back in the deck, and a second card is drawn at random. Find the probability that the first card is a diamond and that the second card is also a diamond.

A. $\dfrac{1}{2}$　　　　　B. $\dfrac{1}{4}$

C. $\dfrac{3}{13}$　　　　D. $\dfrac{1}{16}$

29. A solid box-shaped structure is made of layers of unit cubes stacked one above the other. A $2 \times 3 \times 3$ block of unit cubes has been removed from this structure. Assume that no cubes other than the ones in the region indicated by shading have been removed. How many unit cubes are contained in the structure pictured?

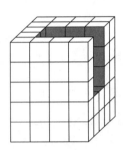

A. 73   B. 82
C. 94   D. 100

30. Rory threw a dart that landed in the 3-point area of the target pictured. Whitney threw a dart that landed in its 1-point area.

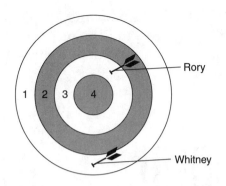

Each of them has just one more dart to throw. It is now Rory's turn. In what area(s) of the target could Rory throw his dart so that he is sure to win; that is, so that Whitney's total points cannot tie or exceed his total points?

A. in the 4-point area only
B. in the 4-point or in the 3-point area only
C. in the 4-point, in the 3-point, or in the 2-point area
D. Rory cannot be sure he will win until after he and Whitney both throw their darts.

PART 2 OPEN-ENDED QUESTIONS

DIRECTIONS: Respond in detail to each open-ended question. Show your process and provide an explanation as directed in the question. Your score for each question will be based on the accuracy and completeness of your process or explanation.

31. The annual salaries of all the employees of a small company are listed below.

President: $320,000
Vice President: $121,000
Senior Professionals: $50,000; $48,000; $48,000; $44,000
Junior Professionals: $36,000; $36,000; $36,000; $32,000
Clerical Staff: $23,000; $22,000

a. What are the mean, the median, and the mode of the salaries of the employees of this company?

b. Which measure of central tendency gives the least accurate picture of salaries at this company? Explain your answer.
c. How will each of these statistics change if the president's salary is excluded?

32. Every Wednesday at the Pizza Express, the manager gives away free slices of pizza and soda. Every eighth customer gets a free slice of pizza and every twelfth customer gets a free soda. The Pizza Express served 87 customers last Wednesday.

a. How many free sodas were given away last Wednesday?
b. How many free slices of pizza were given away?
c. Did any customer receive both a free slice of pizza and a free soda? If so, how many customers?
d. If soda sells for 99¢ and a slice of pizza sells for $1.25, how much did the Pizza Express lose in income by giving away these items? Justify your answer.

33. A standard $8\frac{1}{2}"$ × 11" sheet of paper is rolled along its shorter side to form a cylinder as shown. (Note that the cylinder has no paper on the top or bottom.)

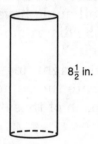

$8\frac{1}{2}$ in.

A second sheet of standard $8\frac{1}{2}"$ × 11" paper is rolled along its longer side to form a second cylinder.

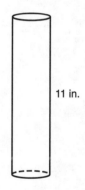

11 in.

There is no overlap of paper on either cylinder.

a. Will the taller cylinder have the same surface area, greater surface area, or less surface area than the shorter cylinder? Explain your answer.
b. Will the taller cylinder have the same volume, greater volume, or less volume than the shorter cylinder? Explain your answer.
c. If a sheet of 11" × 17" paper was used to make a cylinder 17" tall, how would its volume and surface area compare with the volume and surface area of the $8\frac{1}{2}"$ tall cylinder? Explain your answer.

34. A ball was dropped from a height of 16 feet. Each time the ball bounced, it reached a maximum height of approximately half its previous height.

a. Complete the table to show the height reached after the ball bounced each of five times.

Bounce	Height After Bounce
0	Original height = 16'
1	
2	
3	
4	
5	

b. Draw a graph to represent the relationship between the number of the bounce and the height reached by the ball.
c. What is the total vertical distance the ball has traveled after bounce 5? Include the original height of 16' in your total.
d. What do you think the total would have been if the ball had bounced 25 times? Explain your reasoning.

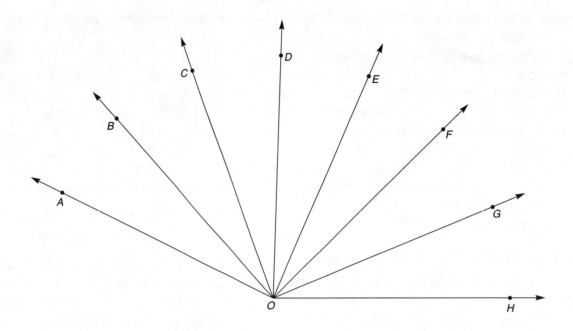

35. The figure shows seven small angles each measuring 22°.

   a. How many obtuse angles are in the figure?

   b. Explain why there is no right angle pictured in the figure.

   c. Suppose you increase the degree measure of each angle slightly, each by the same amount. What would each small angle have to measure so that the diagram would contain right angles?

   d. How many right angles would there be? Indicate the names (using 3 letters) of each of these right angles.

PART 1 MULTIPLE-CHOICE QUESTIONS

1. Which of the solid figures pictured has all three of these characteristics?

   - It has an odd number of faces.
   - It has an even number of vertices.
   - It has more vertices than faces.

A.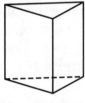

   Triangular Prism

B.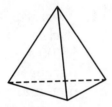

   Triangular Pyramid

C.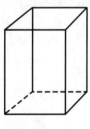

   Rectangular Prism

D.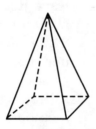

   Rectangular Pyramid

2. Which of the following represents a 5% decrease in price?

   A. $10 to $5
   B. $50 to $45
   C. $100 to $95
   D. $200 to $195

3. Which of these inequalities would be most helpful in solving the following problem?

Bart spent an evening playing video games and drinking sodas. Each video game cost 25 cents to play, and sodas cost 60 cents each. Bart had $8 to spend only on video games and sodas. If he had only 3 sodas and played as many video games as he could, how many video games did he play?

   A. $0.25x + 1.80 \leq 8$
   B. $x + 1.80 \leq 8$
   C. $0.60x + 0.75 \leq 8$
   D. $3x + 0.25 \leq 8$

4. Which statements are equivalent to one another?

   I. Every year, Americans spend $2 million on exercise equipment and $10 million on potato chips.
   II. Every year, Americans spend 5 times as much on potato chips as they do on exercise equipment.
   III. Every year, Americans spend 20% as much on exercise equipment as they do on potato chips.

   A. I and II only   B. I and III only
   C. II and III only   D. I, II, and III

5. *Twin primes* are pairs of prime numbers that are consecutive odd integers, such as 3 and 5, 11 and 13, or 17 and 19. How many pairs of twin primes are there between 10 and 100?

   A. 5   B. 6   C. 7   D. 8

6. If the product of 7 integers is positive, then, at most, how many of the integers could be negative?

A. 2   B. 4   C. 6   D. 7

7. The squares pictured below are all congruent to the one shown at the right. Each of the squares has part of its interior shaded. Which of the squares appears to have the same fraction of its interior shaded as the given one has?

I.   II.   III.   IV.

A. all of them
B. I and II only
C. I, III, and IV only
D. none of them

8. Six slips of paper with the letters A through F written on them are placed into a shoebox. The six slips are drawn one by one from the box. What is the probability that the first three letters drawn are A, D, and F in any order?

A. $\dfrac{1}{20}$ B. $\dfrac{1}{6}$ C. $\dfrac{1}{3}$ D. $\dfrac{1}{2}$

9. Given:
Circle A represents even numbers.
Circle B represents perfect squares.
Circle C represents powers of 10.

Which of the following would be included in the shaded region?

A. {100, 200, 300}
B. {64, 100, 144}
C. {10, 100, 1,000}
D. {100, 10,000, 1,000,000}

10. A particular plant starts as a single stem. At the beginning of the second year, it grows two branches and therefore has three tips growing. Each year, every branch does the same thing; that is, it grows two branches and continues to grow itself. How many tips are there on that plant in the 6th year if no tip or branch has died?

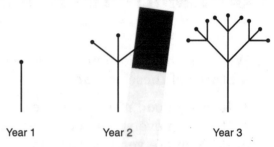

Year 1   Year 2   Year 3

A. 18   B. 81   C. 243   D. 729

11. A printing company makes bumper stickers that cost $0.75 each plus a $5.00 set-up fee. If you spend $80 to purchase a supply of bumper stickers, how many do you get?

A. 50   B. 75   C. 100   D. 150

12. Marah has a $100 gift certificate for Rave Records. After purchasing a DVD for $21.95 and a magazine for $4.95, she would like to buy CD's with her remaining money. What is the greatest number of CD's she can buy if they cost $12.99 each? (Do not include any sales tax.)

A. 5   B. 6   C. 7   D. 8

13. Which of the following would NOT be an illustration of a tessellation?

A. 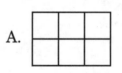   B.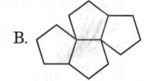

C. 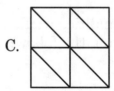   D.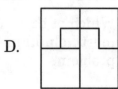

14. During a baseball season, 70 percent of the major league outfielders had at least 20 home runs. Knowing this, determine which of the following must be greater than or equal to 20.

  I. the mean number of home runs
  II. the mode of the number of home runs
  III. the median number of home runs

A. I only   B. II only
C. III only   D. I and II

15. Malcolm graphed all possible combinations of the numbers of correct and incorrect responses students could obtain on a 20-question true-false test. Which graph below MOST LIKELY resembles Malcolm's graph?

A. 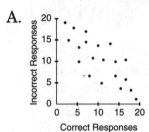   B.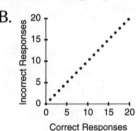

C. 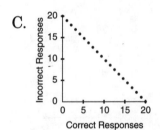   D.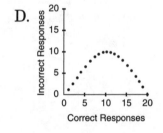

16. How many prime numbers are greater than 90 and less than 100?

A. 0   B. 1   C. 2   D. 3

17. The given figure consists of 9 congruent squares. Which of the following is a square you could remove and not change the perimeter of the entire figure?

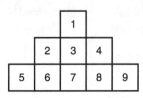

A. square 1   B. square 2
C. square 7   D. square 9

18. How many degrees are in the angle between the hands of a clock at 9:30?

A. 90°   B. 95°
C. 105°   D. 120°

19. Five darts are thrown at the dartboard shown. If they all land in areas pictured (none on the lines), what is the difference between the largest score possible and the smallest score possible?

A. −20   B. 5   C. 25   D. 45

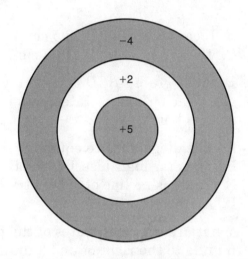

20. A linear function is displayed in the table. When graphed, where would the line cross the x-axis?

x	y
4	3
6	1
10	−3

A. $(0, 7)$ B. $(7, -1)$
C. $(7, 0)$ D. $(8, 0)$

21. Three vertices of an isosceles trapezoid are at $(-4, 0)$, $(4, 0)$, and $(-1, 6)$. The bases of the figure are horizontal. Find the number of square units in the area of the isosceles trapezoid.

A. 27   B. 30   C. 54   D. 60

22. The perimeter of the figure shown is 33 centimeters. If $\overline{BC} \cong \overline{ED}$, what is the measure of $\overline{BC}$?

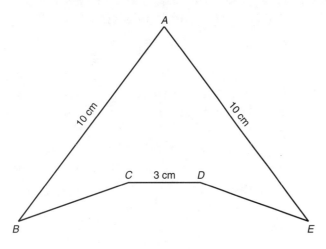

A. 5 cm
B. 5.5 cm
C. 10 cm
D. 11.5 cm

23. Parallelogram $ABCD$ is not a rectangle. Which of these statements is NOT always true?

A. The diagonals are equal in length.
B. The diagonals bisect each other.
C. There is no line of symmetry.
D. All are true.

24. What are the coordinates of the point 6 units to the right of and 3 units below the point $(-7, -1)$?

A. $(6, -3)$
B. $(-10, -7)$
C. $(1, -2)$
D. $(-1, -4)$

25. Which of the following would NOT change the mean for these five scores?

20 30 40 50 60

I. Add two scores: a 10 and a 70.
II. Add three more scores of 40.
III. Add 5 to each score.

A. I only
B. II only
C. I and II
D. I, II, and III

26. The Straight as an Arrow Company paints lines on the streets of small towns. The company charges $100 plus $0.25 per foot. Which of the following is a reasonable graph for length vs. total charge?

A.
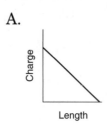

B.
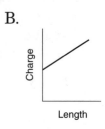

C.
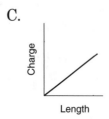

D.
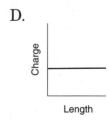

27. To make $\triangle PQR$, congruent to $\triangle ABC$, which of the ordered pairs could be used as the coordinates of point R?

A. $(10, -1)$
B. $(8, 9)$
C. $(10, 5)$
D. $(10, 8)$

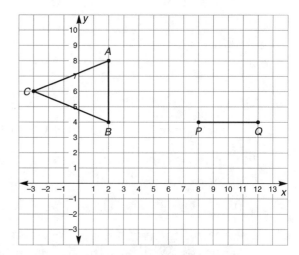

28. If one of the numbers 1, 2, 3, 4, 5, or 6 is substituted randomly into the algebraic expression $3x + 2$, what is the probability that the result will be an odd number?

A. $\dfrac{1}{6}$ B. $\dfrac{1}{3}$ C. $\dfrac{1}{2}$ D. $\dfrac{2}{3}$

29.

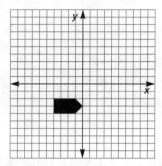

If the figure on the grid above is translated 3 units to the right and then reflected in the *x*-axis, which picture below will show the result?

A.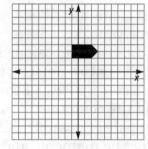

B.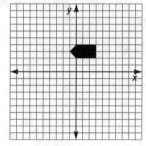

C.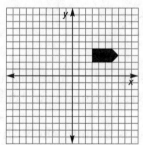

D.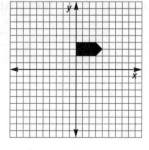

30. The graph below corresponds to Mr. Donatello's auto trip from Boston to Trenton.

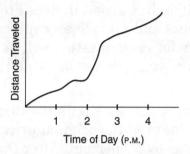

What was most likely happening between 1:45 P.M. and 2:00 P.M.?

A. Mr. Donatello was speeding.
B. Mr. Donatello stopped for a break.
C. Mr. Donatello drove at a steady speed.
D. Mr. Donatello drove in the slow lane.

DIRECTIONS: Respond in detail to each open-ended question. Show your process and provide an explanation as directed in the question. Your score for each question will be based on the accuracy and completeness of your process or explanation.

31. During a sale at Mike's Pet Shop, Mike lowered the original price of iguanas by 30 percent. After the sale, Mike told his clerk to raise the price of iguanas by 30 percent of their sale price. The clerk then marked the iguanas with their original price. Was the clerk right or wrong in doing that? Present a convincing argument to support your answer; you may wish to include a simple, specific example as part of your argument.

32. A cylindrical jar with height 8 inches and diameter 6 inches is filled to 75% of its capacity with juice. The juice is then poured into another cylindrical container with a 10-inch diameter and height of 4 inches.

a. To what percent of its capacity is the second container filled with juice? Show your procedure.

b. A third cylindrical container is such that the entire amount of juice only takes up 27% of the capacity of the container. What would one pair of possible dimensions be for the diameter and height of this third container? Show your approach.

33. Figure *ABCD* is a square.

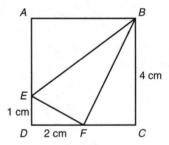

a. To the nearest tenth of a centimeter, find the perimeter of triangle *EFB*.

b. Vanessa was able to show that $\triangle EFB$ was a right triangle. Explain a method to show this.

c. Vanessa found the area of $\triangle EFB$ by using the formula for the area of a triangle. Her friend Matt was able to find the area of $\triangle EFB$ by using the area of other figures. What did Matt do? How many square centimeters are in the area of $\triangle EFB$?

34. Maps are colored so that no two countries (or regions) containing a common border may be the same color. (*Note:* A common border consists of more than a single point.) The following is a map requiring 2 colors:

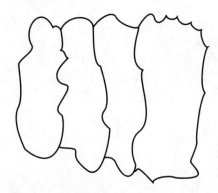

a. Explain why the following map would NOT require 3 colors.

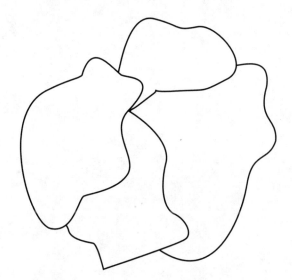

b. Draw a map that would require 3 colors.
c. Draw a map that would require 4 colors.

35. An auditorium has 40 rows of seats. There are 10 seats in the first row, 12 seats in the second row, and so on, with two more seats in each row than in the previous row.

a. Two students, Tom and Brandon, were looking at the seating arrangement in the auditorium. Tom came to the conclusion that 50% of the seats must be in the first 20 rows, since there are 40 rows in the auditorium. Brandon had a feeling that this couldn't possibly be so. Who is correct? Explain your thinking.
b. Tom and Brandon have tickets for seats in row 31. How many seats are in that row? Show how you arrived at your answer.
c. Write an algebraic equation that can be used to find the number of seats in any row of the auditorium. (Use s = the number of seats and r = the row number.)

A

Absolute value, 8
Accuracy, 4
  of measurements, 99
Acute angle, 41
Acute triangle, 48
  area of, 86
Addition
  associative property, 15
  commutative property, 15
  distributive property of multiplication over, 15, 204
  identity property, 15
  inverse property, 15
  of signed numbers, 8
  zero property, 15
Adjacent angles, 41
Algorithm, 166, 173, 177
Angle(s), 41
  acute, 41
  adjacent, 41
  alternate interior, 44
  central, 51
  complementary, 42
  congruent, 64
  corresponding, 44, 64
  interior, 44, 49
  obtuse, 41
  in a polygon, 47
  right, 41
  straight, 41
  supplementary, 42
  vertical, 41
Angle bisector, 41
Approximations, 1–2
Arc, 51
Area, 86–88
  surface, 93–94
Arithmetic operations, properties of, 15
Arithmetic sequences, 190–191
Associative property, 15
Average, 127

B

Bar graph, 132
Base, 26
Best-fit line, 138

C

Capacity (liquid)
  customary conversions for, 97
  metric conversions for, 96

Cartesian coordinate system, 213–214
Center
  of circle, 51–52
  of dilation, 73
Centimeter, 98
Central angle, 51
Central tendency, measures of, 127–128, 141
Charts
  flow, 167–168
  pie, 131
Chord, 51
Circle graph, 131
Circles, 51–52
  area of, 86
  center of, 52
  circumference of, 83–84
  concentric, 52
Circumference, 83–84
Coefficient, 200
Combinations, 153–154
  Pascal's triangle and, 155
Commission, 28
Common difference, 190–191
Common ratio, 192
Commutative property, 15
Complementary angles, 42
Complementary events, 120
Composite number, 17
Compound events, 150
  probability of, 121–123
Compound interest, 163–164
Computational patterns, 185–186
Cone, 56
  volume of, 90
Congruence, 64
Congruent figures, 64
Constant terms, 199
Coordinate plane, 213
Correlation, 138–139
Counting, methods of, 149–156
Counting principle, 150–151, 152
Cross products, 23–24
Cube, volume of, 90
Cubic units, 90
Cup, 97
Customary system of measures, 96–97
Cylinder, 56
  volume of, 90

D

Data
  collection, organization, representation, analysis and evaluation of, 127–128
  displays of, 131–134
  evaluating and interpreting, 140–142
  misleading, 141
  relationships involving, 138–139
Decagon, 47
Decimals, 9
  repeating, 186
Dependent events, 121–122
Diagonal, 47
  of polygon, 40
Diagrams
  iteration, 166–167
  tree, 150–151
Digits, significant, 4–5
Dilation, 72–73
  center of, 73
Discrete mathematics, 149
Distributive property of multiplication over addition or subtraction, 15, 204
Divisible, 17
Division of signed numbers, 9

E

Edge
  of polyhedron, 40
  in network, 159–160
Enlargement, 72–73
Equations, 195, 204, 208
  solving, 204
Equilateral triangle, 48
Estimation, 1–2
  of measurements, 98
  of numbers, 1–2
Euclid, 177
Euler, 159
Events
  complementary, 120
  compound, 150
    probability of, 121–123
  dependent, 121–122
  independent, 121–122, 174
  mutually exclusive, 122–123
  non-mutually exclusive, 122–123
  probability of simple, 117–119
Even vertices, 159
Experimental probability, 119

Exponents, 11–12
  laws of, 11
Extremes, 23

F
Face of polyhedron, 40
Factor, 17
  greatest common, 17
Factorial notation, 151–152
Fair game, 118
Fibonacci sequence, 193
Figures
  congruent, 64
  similar, 66
  three-dimensional, 55–56
  two-dimensional, 47–55
Flow chart, 167–168
Foot, 98
Formulas
  for area, 86
  for compound interest, 163–164
  for perimeter, 83–84
  for solving percent problems,
    26
  for volume, 90
Fractals, 164–165
Fractions, 9
  equivalent, 186
  probability expressed as, 117
Frequency, 127, 132
Functions, 208–211
  non-linear, 210
  step, 196

G
Gallon, 97
Geometric iteration, 162
Geometric relationships, 44–45
Geometric sequence, 192
Geometric terms, 39–42
Gram, 96, 98
Graphs, 140, 159, 195, 208,
  219–220
  bar, 132
  circle, 131
  line, 133
Greatest common factor (GCF),
  17–18, 177

H
Heptagon, 47
Hexagon, 47
Histogram, 132
Hypotenuse, 101

I
Identity property, 15
Image, 69
Inch, 98
Independent events, 121–122,
  174
Inequalities, 205

Initial value, 162
Integers, 8–9, 9
Interest, 163
  compound, 163–164
  simple, 28
Intersection of lines, 44–45
Inverse operation, 204
Inverse property, 15
Isosceles trapezoid, 50
Isosceles triangle, 48
Iteration, 162–163
  geometric, 162
  numerical, 162
Iteration diagram, 166–167

K
Kilogram, 96, 98
Kilometer, 96
Königsberg, bridges of, 159

L
Laws of exponents, 11
Least common multiple (LCM),
  17–18
Length, 98
  customary conversions for, 97
  metric conversions for, 96
Like terms, 200
  combining, 200
Line graph, 133
Line plot, 133
Lines, 39, 44–45
  of best fit, 138
  parallel, 44
  perpendicular, 44
  skew, 45
  slope of, 217–219
  of symmetry, 114
  trend, 138
Line segment, 40
Line symmetry, 69
Liter, 96

M
Magnitude, 8
Mean, 127
Means, 23
Measurements
  accuracy of, 99
  estimating, 98
  systems of, 96–97
Measures of central tendency,
  127–128, 141
Median, 127
Meter, 96, 98
Metric system, 96
  conversions, 96
  prefixes in, 96
Midpoint, 40
Mode, 127
Multiplication
  associative property, 15

commutative property, 15
distributive property of, over
  addition or subtraction, 15, 204
identity property, 15
inverse property, 15
of signed numbers, 8, 9
zero property, 15
Multiple, 17
  least common, 17
Mutually exclusive events,
  122–123

N
Net, 56, 93
Networks, 159–160
N-gon, 47
Nonagon, 47
Non-linear functions, 210
Non-mutually exclusive events,
  122–123
Notation
  factorial, 151–152
  scientific, 13
Not like terms, 200
Number line, 8, 10, 205–206
Numbers
  composite, 17
  fractions, 9
  integers, 8
  opposite, 8
  prime, 17, 20, 168
  rational, 9–10
  sets of, 8–10
  whole, 8, 17
Numerical iteration, 162

O
Obtuse angle, 41
Obtuse triangle, 48
  area of, 86
Octagon, 47
Odd vertices, 159
Open sentences, 201
  translating word problem into,
    201–202
Opposites, 8
Ordered pairs, 208, 213–214
  set of, 195, 208
Order of operations, 19, 199
Orientation, 69
Origin, 213
Ounce, 98
Outcomes, 117–118
Outliers, 138

P
Palindrome, 22
Parallel lines, 44
Parallelogram, 50
  area of, 86
Parentheses, 19, 199
Pascal's triangle, 155–156

Patterns, 185
  computational, 185–186
  numerical, 185–186
  representation of, 194–196
  visual, 187–188
Pentagon, 47
Percent, 26–28
  applications of, 28
  decrease, 28
  increase, 28
  types of problems, 27
Perimeter, 83–84
Permutations, 152–153
Perpendicular lines, 44
Pictograph, 132
Pie chart, 131
Pint, 97
Plane, 40
Point, 39
Polygons, 40, 47
  congruent, 64
  diagonal of, 40
  perimeter of, 83–84
  regular, 47
  similar, 66
  sum of angle measures in, 47
  vertex of, 40
Polyhedron, 40
  edge of, 40
  face of, 40
  vertex of, 40
Population, 130
Pound, 97–98
Powers, 11–12
  of ten, 13, 186
    in scientific notation, 13
Preimage, 69
Prime number, 17, 168
  relatively, 20
Principal, 163
Prism, 55
Probability
  of compound events, 121–123
  experimental, 119
  of simple events, 117–119
  theoretical, 119
Properties of arithmetic operations, 15
  associative, 15
  commutative, 15
  distributive, 15
  identity, 15
  inverse, 15
  zero, 15
Proportion, 23–24
  in similar figures, 66–67
Pyramid, 55
  volume of, 90
Pythagorean Theorem, 101–102

Q
Quadrants, 213

Quadrilaterals, 47, 49–50
Quart, 97

R
Radius, 51
Random sample, 130
Range, 127
Rates, 23–24, 26
  of change, 216–217
  unit, 24
Ratio, 23–24
  common, 192
  of similitude, 66
Rational numbers, 9–10
Ray, 39
Reasonableness of results, 3
Rectangle, 50
  area of, 86
  perimeter of, 83–84
Rectangular prism, volume of, 90
Recursion, 162–163
Reduction, 72–73
Reflection, 69
Regular polygon, 47
Relations, 208
Relationships
  picturing, 219–220
  representation of, 194–196
Relatively prime numbers, 20
Rhombus, 50
Right angle, 41
Right triangle, 48
  area of, 86
Roots, 12
  cube, 12
  square, 12
Rotation, 70–71
Rounding, 1

S
Sample, 130
  random, 130
  unbiased, 130
Sample space, 117
Sampling, 130
Scalene triangle, 48
Scatter plot, 138–139
Scientific notation, 13
Seed value, 162
Self-similarity, 164
Sentences, open, 201–202
Sequences, 190
  arithmetic, 190–191
  Fibonacci, 193
  geometric, 192
Set
  of numbers, 8–10
  of ordered pairs, 195, 208
Sierpinski triangle, 164
Sieve of Eratosthenes, 168
Signed numbers
  addition of, 8

  division of, 9
  multiplication of, 9
  subtraction of, 9
Significant digits, 4–5
Similar figures, 66
Similarity, 66
  in dilation, 72
Simple events, probability of,
  117–119
Skew lines, 45
Slope, 217–219
Sphere, 56
  volume of, 90
Spreadsheets, 134–135
Square, 50
  area of, 86
  perimeter of, 83–84
Square root, 12
Statistical measures, 127–128, 141
Statistical study, 130
Step function, 196
Straight angle, 41
Subtraction
  distributive property of multipli-
    cation over, 15, 204
  of signed numbers, 9
Supplementary angles, 42
Surface area, 93–94
Symmetry, line, 69

T
Table of values, 195, 216–217
Tables, 141, 208
Tax, sales, 28
Temperature, 101
Terms, 199
Tessellation, 76–77
Theoretical probability, 119
Three-dimensional figures, 55–56
Ton, 97
Transformations, 69–73
Translation, 71–72
Transversal, 44
Trapezoid, 50
  area of, 86
  isosceles, 50
Tree diagram, 150–151
Trend line, 138
Triangles, 47, 48
  acute, 48
  classification of, 48
  equilateral, 48
  isosceles, 48
  obtuse, 48
  Pascal's, 155–156
  right, 48
  scalene, 48
Two-dimensional figures, 47–55

U
Unbiased sample, 130
Unit rate, finding, 24

V
Variable, 199
Variable expressions, 199
  translating from verbal statements to, 201
Verbal statement(s), 194
  translating to variable expressions, 201
Vertical angles, 41
Vertex (vertices)
  even, 159
  in network, 159–160
  odd, 159

of polygon, 40
of polyhedron, 40
Visual patterns, 187–188
Volume, 90–91

W
Weight (mass)
  customary conversions for, 97
  metric conversions for, 96
Whole number, 17
Word problem, translating into
  open sentence, 201–2

X
x-axis, 213
x-coordinate, 214

Y
Yard, 98
y-axis, 213
y-coordinate, 214

Z
Zero, significance of, 5
Zero property, 15